POLED POLYMERS
AND THEIR APPLICATIONS
TO SHG AND EO DEVICES

Advances in Nonlinear Optics
A series edited by Anthony F. Garito, *University of Pennsylvania, USA*
and François Kajzar, *DEIN, CEN de Saclay*, France

This book is part of a series. The publisher will accept continuation orders which may be cancelled at any time and which provide for automatic billing and shipping of each title in the series upon publication. Please write for details.

POLED POLYMERS AND THEIR APPLICATIONS TO SHG AND EO DEVICES

Edited by

Seizo Miyata
Tokyo University of Agriculture and Technology
Japan

and

Hiroyuki Sasabe
Frontier Research Program
RIKEN
Japan

TAYLOR & FRANCIS
ALERE FLAMMAM
Founded 1798

First published 1997 by Gordon and Breach
Second printing 2001

Transferred to Digital Printing 2002
by Taylor & Francis
11 New Fetter Lane, London EC4P 4EE

Taylor & Francis is an imprint of the Taylor & Francis Group

British Library Cataloguing in Publication Data

Poled polymers and their applications to SHG and EO devices. – (Advances in nonlinear optics; v. 4)
   1. Polymers – Electric properties   2. Polymers – Magnetic properties   3. Organic electrochemistry   4. Electro-optics   5. Molecular electronics
   I. Sasabe, Hiroyuki   II. Miyata, Seizo, 1941-
   547.7'04578

ISBN 90-5699-025-X

CONTENTS

INTRODUCTION TO THE SERIES

Advance in Nonlinear Optics is a series of original monographs and collections of review papers written by leading specialists in quantum electronics. It covers recent developments in different aspects of the subject including fundamentals of nonlinear optics, nonlinear optical materials both organic and inorganic, nonlinear optical phenomena such as phase conjugation, harmonic generation, optical bistability, fast and ultrafast processes, waveguided nonlinear optics, nonlinear magneto-optics and waveguiding integrated devices.

The series will complement the international journal *Nonlinear Optics: Principles, Materials, Phenomena and Devices* and is foreseen as material for teaching graduate and undergraduate students, for people working in the field of nonlinear optics, for device engineers, for people interested in a special area of nonlinear optics, and for newcomers.

PREFACE

This monograph on *Poled Polymers and their Applications to SHG and EO Devices* contains papers presented at the Satellite Symposium of the *2nd International Conference on Organic Nonlinear Optics (ICONO'2)*, which was held at Hotel Village, Kusatsu, Gumma, Japan, July 23–26, 1995.

Poled polymers doped with nonlinear optically active chromophores were demonstrated to combine the large second-order nonlinearity of the dopant dye molecules with the optical quality of the polymer. Recently, much progress has been made in making highly stable poled polymeric materials. Basic studies to shed light on the relaxation mechanisms in polymers, aimed at further improving material stability and reliability have seen recent advances. The material design flexibility afforded to doped polymers makes them attractive in a large variety of devices and applications.

The symposium addressed critical science and technology issues in the development and application of poled polymers by providing a forum for international collaboration. Emphasis was placed on the stabilization of poled polymers and special applications to second harmonic generation (SHG) and electro-optic (EO) devices. Participants of this symposium numbered 159 from 16 different countries, and enjoyed not only mutual discussion on newly emerging technologies in photonics materials and their promising device applications, but also a scenic view and hot spring as well.

We would like to express our gratitude to all members of the organizing committee and the scientific program committee of the symposium for their help in establishing an excellent program. We thank Drs S. Fujishiro and S. J. Yakura of the Asian Office of Aerospace Research and Development (AOARD), Dr Charles Y.-C. Lee of the Airforce Office of Scientific Research (AFOSR), Dr K. J. Wynne of the Office of Naval Research (ONR) and Dr I. Ahmad of the Army Research Office (ARO) for their financial support and invaluable assistance in all phases of the organizing process. The sponsorship of AOARD towards publishing this monograph is also highly appreciated.

Seizo Miyata
Chairman of ICONO'2 and Editor

Hiroyuki Sasabe
Secretary General of ICONO'2
and Editor

1. POLED POLYMERS AND THEIR APPLICATION IN SECOND HARMONIC GENERATION AND ELECTRO-OPTIC MODULATION DEVICES

F. KAJZAR and P.-A. CHOLLET

CEA (LETI-Technologies Avancées), DEIN/SPE, Groupe Composants Organiques, CEA Saclay, 91191 Gif sur Yvette Cedex, France

ABSTRACT

Different aspects of molecular engineering, from second order nonlinear optical response characterization on molecular level, oriented thin film preparation techniques and characterization of their linear and nonlinear optical properties are described and discussed. A particular attention is paid to poled polymers which offer good prospects for practical applications such as frequency doubling and electro-optic modulation, both in the waveguiding configuration. New poling techniques, such as photo-assisted and all optical poling are also described and discussed. Possible applications of the noncentrosymmetric thin films for electro-optic modulation and frequency conversion in waveguiding configuration are also reviewed and discussed.

1 INTRODUCTION

It is generally admitted that the next century will be a century of information. It will require the transmission of a large volume of information between sources and users, especially as interactive links are envisaged. Presently optical fibers offer a large capacity information transmission, with small losses. The problems exist at the interface between source/optical fiber and optical fiber/user interfaces, which at present are electronic. This limits strongly the amount of information which can be sent as well as the speed, because of the use of electrical circuitry. Similar problems exist with the further development of computers where limitations arise from interconnections. Here optical interconnections are expected to bring a big improvement.

Another encountered problem is the information storage capability. Optical discs offer the largest storage capability. At the present time the information is written and read with laser diode emitting at 830 nm. The storage capacity is limited by the resolution. By dividing the writing/reading light source wavelength by two one multiplies the storage capacity on the same surface by a factor of four. There exists also an increasing need for laser sources emitting in blue, for biological, medical or electronic applications (photolithography). All types of applications discussed here can be made uniquely with noncentrosymmetric materials, although a second order NLO response has been also found in centrosymmetric materials (Chollet *et al.*, 1991; Hoshi *et al.*, 1991; Wang *et al.*, 1992; Kajzar *et al.*, 1993; Koopmans *et al.*, 1993; Yamada *et al.*, 1996). These nonlinearities are too small to imagine realistic practical applications. It represents, however, a scientific interest. For the forthcoming applications the main emphasis is done presently on electro-optic modulation and frequency doubling devices.

Much work have been done with elaboration of single crystals, mainly for frequency conversion (for a historical review see Kajzar and Zyss, 1995b). At the present time one can find on the market frequency doublers, triplers, or parametric oscillators, made principally from inorganic single crystals. Very interesting opportunities represent organic molecules, which exhibit large first hyperpolarizability β coefficients, offering a great potential for this kind of applications. However for device applications important is the macroscopic second order nonlinear optical susceptibility $\chi^{(2)}$, which depends not only on the molecular first hyperpolarizability β value but on the way how molecules have been assembled into active, bulk material. Indeed, in dipolar approximation, the molecular polarization can be expanded into the forcing field series and its ith component is given by the following expression:

$$p_i = p_{0i} + \varepsilon_0(\alpha_{ij}E_j + \beta_{ijk}E_jE_k + \gamma_{ijkl}E_jE_kE_l + \cdots), \tag{1}$$

where the development coefficients α_{ij}, β_{ijk}, γ_{ijkl} are three dimensional, 2nd, 3rd and 4th rank tensors respectively, describing the molecular response under exciting local fields E, experienced by molecule. For centrosymmetric molecules it follows from the symmetry considerations and within the dipolar approximation that the first hyperpolarizability β tensor has all components equal to zero. There is also no static polarization on the molecule ($\mathbf{p}_0 = \mathbf{0}$).

Similarly, for a bulk material and on the macroscopic level the medium polarization can be also expanded into the external forcing field power series giving the following expression, again in the electric dipolar approximation:

$$P_i = P_{0i} + \varepsilon_0(\chi_{ij}^{(1)}E_j + \chi_{ijk}^{(2)}E_jE_k + \chi_{ijkl}^{(3)}E_jE_kE_l + \cdots), \tag{2}$$

where the development coefficients $\chi^{(n)}$ ($n = 1, 2, 3, \ldots$) are three dimensional, $(n + 1)$ rank tensors describing the medium response to the exciting external fields E. Similarly as before for centrosymmetric materials and within the dipolar approximation all odd rank tensors, as well as the static polarization $\mathbf{P}_0$ are equal to zero.

There exists a simple relationship between macroscopic and microscopic hyperpolarizabilities in the case of a long range, three dimensional order (as it is the case for e.g. single crystals). This is obtained by doing a transformation of the corresponding quantities from the molecular reference frame (x, y, z) to the laboratory one: (X, Y, Z). This can be done using the Wigner's rotation matrices a_{ij}. For different linear and nonlinear optical susceptibility tensor components one gets the following relationships between macroscopic and the corresponding microscopic quantities:

$$\chi_{IJ}^{(1)}(-\omega; \omega) = N \sum_{ijk} f_i^\omega f_j^\omega a_{iI} a_{jJ} \alpha_{ij} \tag{3}$$

for linear suceptibilitiy,

$$\chi_{IJK}^{(2)}(-\omega_3; \omega_1, \omega_2) = N \sum_{ijk} f_i^{\omega_1} f_j^{\omega_2} f_k^{\omega_3} a_{iI} a_{jJ} a_{kK} \beta_{ijk}(-\omega_3; \omega_1, \omega_2) \tag{4}$$

for second order nonlinear optical susceptibility and

$$\chi^{(3)}_{IJKL}(-\omega_4; \omega_1, \omega_2, \omega_3) = N \sum_{ijkl} f^{\omega_1}_i f^{\omega_2}_j f^{\omega_3}_k f^{\omega_4}_l a_{iI} a_{jJ} a_{kK} a_{lL} \gamma_{ijkl}(-\omega_4; \omega_1, \omega_2, \omega_3) \quad (5)$$

for third order nonlinear optical susceptibility.

In Eqs. (3)–(5) N is the number density of molecules, a_{IJ}'s are Wigner's rotation matrix elements allowing a transformation of the molecular reference frame (ijk) to the laboratory system (IJK). f's are the usual local field factors. If there is the same density of molecules along the three directions I, J and K and for optical frequencies these local field factors are given by the well known Lorentz–Lorenz formula:

$$f^i_\omega = \frac{n^2_{(i)\omega} + 3}{2}, \quad (6)$$

where $n_{(i)\omega}$ is index of refraction in direction (I).

In a general case, when no long range order is present, one can introduce an orientation distribution function of the molecules $G(\Theta, \Phi, \Psi)$. The corresponding macroscopic susceptibilities are given by the following equations:

$$\chi^{(1)}_{IJ}(-\omega; \omega) = N \sum_{ij} f^\omega_i f^\omega_j \langle a_{iI} a_{jJ} | G(\Theta, \Phi, \Psi) \rangle \alpha_{ij}(-\omega; \omega), \quad (7)$$

$$\chi^{(2)}_{IJK}(-\omega_3; \omega_1, \omega_2) = N \sum_{ijk} f^{\omega_1}_i f^{\omega_2}_j f^{\omega_3}_k \langle a_{iI} a_{jJ} a_{kK} | G(\Theta, \Phi, \Psi) \rangle \beta_{ij}(-\omega_3; \omega_1, \omega_2), \quad (8)$$

$$\chi^{(3)}_{IJKL} = N \sum_{ijkl} f^{\omega_1}_i f^{\omega_2}_j f^{\omega_3}_k f^{\omega_4}_l \langle a_{iI} a_{jJ} a_{kK} a_{lL} | G(\Theta, \Phi, \Psi) \rangle \gamma_{ijk}(-\omega_4; \omega_1, \omega_2, \omega_3). \quad (9)$$

As already mentioned, in this paper we will limit ourselves up to second order nonlinear optical properties, described by the nonlinear optical susceptibility $\chi^{(2)}_{IJK}$. The relationships between microscopic and macroscopic quantities for different crystal structures can be found in the paper by Zyss and Oudar (1982). The recent work of Kajzar *et al.* (1995a) shows how combining solution EFISH with single crystal SHG and electro-optic coefficient measurements and knowing single crystal structure, using a resonant two level model one can get within a given set of conventions a homogenous description of different second order nonlinear optical effects as well as a correct relationship between macroscopic and microscopic quantities.

Theoretical calculations may give a hint concerning the design of suitable molecules with enhanced first hyperpolarizability β tensor. The molecules can be further synthesized by an organic chemist. Both can thus intervene at the molecular level through so called molecular engineering. However for practical applications the macroscopic second order nonlinear optical susceptibility $\chi^{(2)}$ is of importance. As already mentioned and as it show Eqs. (3)–(4) this susceptibility depends not only the molecular hyperpolarizability β value but also on the way how the molecules have been assembled to build up the bulk material. In the extreme case of macroscopically isotropic thin films we can have molecules with enhanced first hyperpolarizability leading to zero second order nonlinear optical susceptibility $\chi^{(2)}$. Thus a great deal of research is going on towards optimization of ordered materials (or thin films) built up with enhanced second order nonlinear optical susceptibility $\chi^{(2)}$ by an optimized molecular assembly.

Equations (4) and (8) show that the relation between microscopic and macroscopic hyperpolarizabilities is quite complicate as there are in a general case, 27 independent components of the β and $\chi^{(2)}$ tensors. It becomes relatively simple for one dimensional charge transfer molecules, for which there is an enhanced component β_{zzz} of the first hyperpolarizability tensor β, component in the charge transfer direction z which is also the direction of the dipole moment μ (Figure 1). In that case all other components can be neglected with respect to this one.

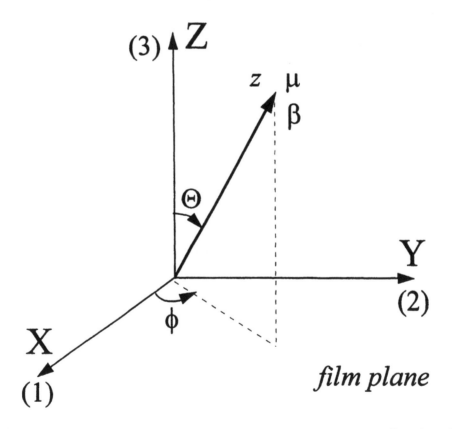

Figure 1 Orientation of the charge transfer molecules with respect to the film plane. The dominant β-component and the dipole moment μ are assumed to be oriented in the same direction.

Different types of applications are targetted with noncentrosymmetric thin films. These include (cf. Kajzar and Swalen, 1996):

— Frequency doubling for data storage, microlithography, biological applications
— Electro-optic modulation for applications in telecommunications, optical inter-connects
— Parametric conversion for fabrication of tunable laser sources
— Sensor applications.

All these applications are targetted in waveguiding configuration which offers an unique opportunity of getting higher optical fields through important field confine-ment. It gives also a possibility of integration of nonlinear optical element with semiconductor technology. The goal is to get detector, active element and light source (laser diode) on the same chip.

There exist several nonlinear waveguided optics requirements which has to be fulfilled, such as:

— Easy thin film fabrication
— Possibility of integration with semiconductor technology
— Good optical quality (propagation losses < dB/cm)
— Large value of $\chi^{(2)}$ as efficiency is proportional to $(\chi^{(2)}l)^2$ and the propagation length l is usually limited by the optical quality and the device size.

Waveguided nonlinear optics offers several important advantages such as:

— Electric field confinement leading to large electric fields E within waveguide
— (Efficiency is proportional to E^4 in the case of frequency conversion or to E^2 for electro-optic modulation)
— Integrated devices
— Electro-optical signal processing with low AC fields for signal modulation.

Depending on targetted applications different criteria rule the choice of active molecules. These include:

— Transparency range. For "blue conversion" (from near IR to UV) one needs molecules with an absorption band cut off around 300–350 nm, whereas for e.g. electro-optic modulation in telecommunication range these may absorb in visible.
— Poling efficiency ($\sim \mu E_p/kT$), which depends on the molecule permanent dipolar moment, poling field E_p and poling temperature T. Poling fields are limited by dielectric breakdown, whereas the molecule dipolar moment depends on design of the molecule. The charge transfer molecules have usually large permanent dipolar moment
— Photochemical and thermal stability. This is an important point which attracts more and more attention as some molecules undergo photo- or thermally induced degradation during operation.

In this paper we review recently developed oriented thin film fabrication tech-niques for second order nonlinear optics. We describe shortly linear and nonlinear optical properties characterization techniques as well as device developments.

2 MOLECULAR FIRST HYPERPOLARIZABILITY DETERMINATION METHODS

2.1 Solvatochromism

Several techniques for determination of molecular hyperpolarizabilities of dipolar molecules have been developed. The simplest one is the solvatochromism (Liptay, 1969; Paley et al., 1989; Gonin et al., 1994). Depending on the solvent polarity (or in other term on the electric field experienced by molecule) one observes a more or less pronounced shift of the molecule absorption spectrum. The difference between vacuum v_m and solution $(v_m)_s$ maximum absorption wavenumber is given by

$$(v_m)_s - v_m = K\frac{n^2-1}{2n^2+1} + L\left(\frac{\varepsilon-1}{\varepsilon+2} - \frac{n^2-1}{n^2+2}\right) + \text{specific interactions} \qquad (10)$$

with

$$L = \frac{2}{cha}\mu_{00}(\mu_{00} - \mu_{11}) \qquad (11)$$

where a is the radius of the spherical cavity occupied by the molecule in the solvent, ε is solution static dielectric constant and n its refractive index, μ_{00} and μ_{11} are permanent dipolar moments at fundamental (0) and excited (1) state. The fundamental state dipolar moment is determined by independent measurements (for details see e.g. Gonin et al. (1994)). The molecular first hyperpolarizability is obtained from two level model (Oudar and Chemla, 1975) through the following formulas (Gonin et al., 1994):

$$\beta(-2\omega;\omega,\omega) = \frac{1}{6\hbar^2}|\mu_{01}|^2 F(\omega,\omega_0,\Gamma), \qquad (12)$$

where

$$F(\omega,\omega_0,\Gamma) = \frac{1}{(\Omega-2\omega)(\Omega-\omega)} + \frac{1}{(\Omega^*+2\omega)(\Omega^*+\omega)} + \frac{1}{(\Omega^*+\omega)(\Omega-\omega)} \qquad (13)$$

describes dispersion of first hyperpolarizability with $\Omega = \omega_0 - i\Gamma$ and $\Omega^* = \omega_0 + i\Gamma$, where ω_0 is transition energy between fundamental and first (allowed) state and Γ is the damping term. The dipolar transition moment can be determined from the optical absorption spectrum using the following relation:

$$\int_{\text{BAND}} \frac{\varepsilon^s(v)}{(v_m)_s} dv = \frac{8\pi^3 N_A}{3hc\ln(10)}|\mu_{01}|^2, \qquad (14)$$

where N_A is the Avogadro number and as previously $(v_m)_s$ is the maximum absorption wavenumber. Equations (10)–(13), assuming the ground state dipolar moment has been determined (from the dielectric constant of the solution), and the transition moment from the absorption spectrum (cf. Eq. 14) yields the molecular first hyperpolarizability β. However, as discussed e.g. by Gonin *et al.* (1994) this technique is not very precise and the result depends on the class of solvents used, due to specific interactions between solvent and solute, leading to e.g. hydrogen bond creation.

2.2 Electric Field Induced Second Harmonic Generation (EFISH)

The second common method for molecular first hyperpolarizability determination is the electric field induced second harmonic generation (EFISH) technique in solution (cf. Levine and Bethea, 1975; Oudar, 1977; Singer and Garito, 1981; Kajzar *et al.*, 1987). The technique can be applied only to dipolar molecules. Under an applied external electric field E molecules in solution orient almost in its direction giving rise to second harmonic generation. The measured third order nonlinear optical susceptibility is given by the following expression:

$$\chi^{(3)}(-2\omega;\omega,\omega,0) = N(f^\omega)^2 f^{2\omega}\left[f^0\gamma(-2\omega;\omega,\omega,0) + \frac{\mu_z\beta_z}{5kT}\right]E, \qquad (15)$$

where z is direction of molecule dipolar moment (usually the direction of the charge transfer axis) and

$$\gamma(-2\omega;\omega,\omega,0) = \tfrac{1}{5}(\gamma_{xxxx} + \gamma_{yyyy} + \gamma_{zzzz} + 2[\gamma_{xxyy} + \gamma_{xxzz} + \gamma_{yyzz}]) \qquad (16)$$

and the vector part of first hyperpolarizability tensor is given by

$$\beta_z(-2\omega;\omega,\omega) = \beta_{zzz} + \beta_{zxx} + \beta_{zyy}. \qquad (17)$$

The second hyperpolarizability $\gamma^{(2)}(-2\omega;\omega,\omega,0)$ on RHS of Eq. (15) is usually neglected with respect to the orientational term. This neglect is almost justified in the case of small, weakly conjugated molecules. As $\gamma(-2\omega;\omega,\omega,0)$ increases much stronger with conjugation length than $\beta(-2\omega;\omega,\omega)$ this assumption is not satisfied in the case of more conjugated molecules. In that case $\gamma(-2\omega;\omega,\omega,0)$ has to be determined by other technique, e.g. third harmonic generation (THG). As a matter of fact THG technique gives $\gamma(-3\omega;\omega,\omega,\omega)$ hyperpolarizability which differs from $\gamma(-2\omega;\omega,\omega,0)$ due to conventions and different dispersion law . A correct approach consists of a measurement of $\gamma(-3\omega;\omega,\omega,\omega)$ spectrum. By applying essential state model one determines different dipolar transition moments as well as differences of dipolar moments between fundamental and excited states in the case of non-centrosymmetric molecules. These can be thus used to calculate $\chi^{(3)}(-2\omega;\omega,\omega,0)$ at given (measurement) frequency (Messier *et al.*, 1991; Burland *et al.*, 1991; Gonin *et al.*, 1994).

2.3 Harmonic Light Scattering

Harmonic Light Scattering (HLS) (Cyvin, 1965; Terhune, 1970; Clays, 1991; 1992), called also Hyper-Rayleigh Scattering (HRS, Clays, 1991; 1992; Verbiest, *et al.*, 1992) is a relatively simple technique allowing different components of first hyperpolarizability determination without molecular orientation, thus no permanent dipole requirement. The measurements are done in solution and the scattered light intensity in y-direction for the z-polarized incoming beam is given by:

$$I_z^{2\omega} = g[N_s \langle \beta_{zzz}^2 \rangle_s + N_p \langle \beta_{zzz}^2 \rangle_p] e^{-\varepsilon(2\varepsilon)lN_p} I_\omega^2, \qquad (18)$$

$$I_x^{2\omega} = g[N_s \langle \beta_{zxx}^2 \rangle_s + N_p \langle \beta_{zxx}^2 \rangle_p] e^{-\varepsilon(2\omega)lN_p} I_\omega^2, \qquad (19)$$

where $N_{s,p}$ are number densities of the solvent (s) and solute (p), respectively, $\varepsilon(2\omega)$ the molar extinction coefficient of the polymer at 2ω frequency and l an effective optical path. The factor g in Eqs. (18)–(19) takes account of local fields and geometrical factors and the averages are given by:

$$\langle \beta_{zzz}^2 \rangle = \frac{1}{7}\sum_i \beta_{iii}^2 + \frac{6}{35}\sum_{i \neq j}\beta_{iii}\beta_{ijj} + \frac{9}{35}\sum_{i \neq j}\beta_{ijj}^2 + \frac{6}{35}\sum_{i,j,k,\text{cyclic}}\beta_{iij}\beta_{jkk} + \frac{12}{35}\beta_{ijk}^2 \qquad (20)$$

and

$$\langle \beta_{zxx}^2 \rangle = \frac{1}{35}\sum_i \beta_{iii}^2 - \frac{2}{105}\sum_{i \neq j}\beta_{iii}\beta_{ijj} + \frac{11}{105}\sum_{i \neq j}\beta_{ijj}^2 - \frac{2}{105}\sum_{i,j,k,\text{cyclic}}\beta_{iij}\beta_{jkk} + \frac{8}{35}\beta_{ijk}^2. \qquad (21)$$

Although, *a priori*, there is a large number of tensor components intervening in Eqs. (20)–(21) for limited number of experimental data, their number is usually limited by molecular symmetry to a smaller number, of which only some are relevant. For example: in quasi 1D charge transfer molecules like para-nitroalinine the tensor component in the charge transfer axis is strongly enhanced and the others can be neglected with respect to this one.

The tensor components are obtained from the slope of the scattered light intensity at harmonic frequency *versus* the square of the fundamental beam intensity, measured for different polarization configurations. The value of g parameter, intervening in Eqs. (19)–(20) is obtained by calibration with solvent with known, β-value. These measurements have to be done very carefully, as another processes, like two photon induced luminescence, Raman and higher order NLO processes can contribute to the measured HLS signal.

Provided that all these effects are correctly taken into account, the technique is simple, as already mentioned and can be applied to molecules without a permanent dipole moment at fundamental state, like octupolar molecules (Zyss, 1991).

2.4 Solid Solutions

The first hyperpolarizability tensor for dipolar molecules can be also measured in solid solution. By applying an external DC field (poling field) one orients chromophores in its direction. For a moderately poled film the two nonzero second order NLO susceptibility tensor components are given by the following expressions:

$$\chi^{(2)}_{333}(-2\omega;\omega,\omega) = N(f^\omega)^2 f^{2\omega}\left[\gamma(-2\omega;\omega,\omega,0)f^0 E_p + \beta_{zzz}L_3\left(\frac{f^0\mu E_p}{kT_p}\right)\right] \quad (22)$$

where L_3 is the Langevin function given by

$$L_3(x) = \int_0^\pi e^{-x\cos\theta}\cos^3\theta\sin\theta\,d\theta \bigg/ \int_0^\pi e^{-x\cos\theta}\sin\theta\,d\theta \quad (23)$$

and f^0 is the Onsager local field factor given by:

$$f^0 = \frac{\varepsilon_{st}(n_\omega^2 + 2)}{n_\omega^2 + 2\varepsilon_{st}}. \quad (24)$$

In Eq. (22) E_p and T_p are the poling fields and temperature respectively, and similarly as before f^i's are local field factors at frequency i. $\chi^{(2)}_{113}(-2\omega;\omega,\omega)$ is given by replacing in Eq. (22) L_3 by:

$$L_1(x) = \int_0^\pi e^{-x\cos\theta}\cos\theta\sin^3\theta\,d\theta \bigg/ \int_0^\pi e^{-x\cos\theta}\sin\theta\,d\theta. \quad (25)$$

The second hyperpolarizability $\gamma(-2\omega;\omega,\omega,0)$ may be for instance obtained from THG measurements on unpoled films (yielding $\gamma(-3\omega;\omega,\omega,\omega)$ and a subsequent use of a two or three level model. If the poling field E_p as well as the number density N are known one can determine from SHG measurements the first hyperpolarizability $\beta(-2\omega;\omega,\omega)$. We note that a systematic neglect of $\gamma(-2\omega;\omega,\omega,0)$ term in Eq. (23) is not justified, especially in the case of slightly doped films and a weak poling. For $\mu E_p/kT_p < 1$ and small $\gamma(-2\omega;\omega,\omega,0)$ term with respect to the second term on RHS of Eq. (22) one obtains a simplified formula

$$\chi^{(2)}_{333}(-2\omega;\omega,\omega) = N(f^\omega)^2 f^{2\omega}f^0\beta_{zzz}\frac{\mu E_p}{kT_p}. \quad (26)$$

The electro-optic first hyperpolarizability, $\beta(-\omega,\omega,0)$ can be obtained from the electro-optic $\chi^{(2)}_{333}$ susceptibility mesurements on, e.g. poled films (see Kajzar et al., 1994) using the following relation:

$$\chi^{(2)}_{333}(-\omega;\omega,0) = N(f^\omega)^2 f\left[\gamma(-\omega;\omega,0,0)f^0 E_p + \beta_{zzz}(-\omega;\omega,0)L_3\left(\frac{f^0\mu E_p}{kT_p}\right)\right], \quad (27)$$

where similarly as before L_3 is the Langevin function, E_p and T_p are poling fields and temperature, respectively, and f^i's are local field factors at frequency i. For low poling fields and small $\gamma(-\omega;\omega,0,0)$ this equation can be simplified giving $\chi^{(2)}(-\omega;\omega,0)$ according to

$$\chi^2_{333}(-\omega;\omega,0) = N(f^\omega)^2(f^0)^2 \beta_{zzz}\frac{\mu E_p}{kT_p}. \tag{28}$$

The SHG first hyperpolarisability $\beta(-2\omega;\omega,\omega)$ can be thus obtained by use of the two level model. Within this model the electro-optic first hyperpolarizability is given by

$$\beta(-\omega;\omega,0) = \frac{1}{3\hbar^2}|\mu_{01}|^2 F_{EO}(\omega,\omega_0,\Gamma), \tag{29}$$

where

$$F_{EO}(\omega,\omega_0,\Gamma) = \frac{1}{(\Omega-\omega)^2} + \frac{1}{(\Omega-\omega)\Omega^*} + \frac{1}{(\Omega^*+\omega)\Omega^*}$$
$$+ \frac{1}{\Omega(\Omega-\omega)} + \frac{1}{\Omega(\Omega^*+\omega)} + \frac{1}{(\Omega^*+\omega)^2} \tag{30}$$

with similar as before meaning for Ω and Ω^* (cf. Eq. (13)).

3   ORIENTED THIN FILM FABRICATION TECHNIQUES

Several techniques have been developed for the fabrication of noncentrosymmetric thin films, including:
 (i) Langmuir–Blodgett (LB) technique
 (ii) Epitaxy or heteroepitaxy
 (iii) Static field poling
 (iv) Photoassisted poling
 (v) All optical poling.

3.1 Langmuir–Blodgett Technique

The LB techniques requires specially designed molecules, containing a charge transfer NLO element and amphiphillic groups, with hydrophobic and hydrophilic terminal groups, respectively. First observation of SHG from an LB monolayer was reported by Aktsipetrov et al. (1983). Although the technique leads to ordered, with well controlled thickness thin films, usually the dipolar molecules can be deposited in Y type structure only, instead of desired for quadratic NLO X or Z type (cf. Figure 2) In order to overcome these difficulties the use of alternate layers has been proposed. However even in that case it is difficult to built up sufficiently thick films,

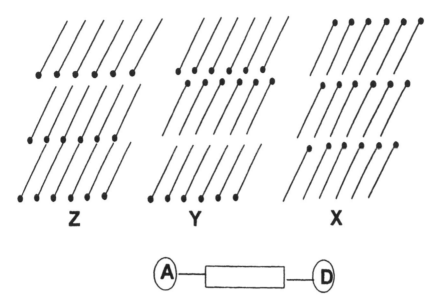

Figure 2 Schematic representation of different Langmuir–Blodgett build-up structures and chemical structure of the optically nonlinear charge transfer molecule with the electron donor (D) and acceptor (A) groups connected by a delocalized π-electron system.

with acceptable propagation losses (Bosshard *et al.*, 1996). As a matter of fact LB films exhibit important propagation losses due to the light scattering by crystallites. Only a few successful results have been reported with LB films (Bosshard *et al.*, 1991). LB films with acceptable losses were obtained using DCANP molecule which orients during the thin film transfer with two nonzero d tensor components: in the drawing direction and perpendicularly to it. They demonstrated an efficient Cerenkov second harmonic generation from a waveguide. Better results, concerning propagation properties, have been obtained with LB films made from preformed polymers (Penner *et al.*, 1994). Despite the above mentioned problems the LB films appear to be useful for studying, e.g. the spectral response of molecules through nonlinear spectroscopy (Kajzar and Ledoux, 1989; Shen, 1989), where one takes advantage of their uniformity and well controlled thickness.

3.2 Epitaxy and Heteroepitaxy

Epitaxy requires the use of a monocrystalline substrate with a well adapted crystal structure with respect to that of the epitaxied molecules. It is necessary that the contact plane of the substrate matches well with the epitaxied molecule crystallographic edges (or their rational multiples) in two directions. A successful molecular epitaxy has been demonstrated with a polydiacetylene (*p*-DCH) (Le Moigne *et al.*, 1991) using a single crystal of phthalate acid of potassium (KAP) as substrate. Monomer thin films have been deposited by vacuum sublimation and subsequently polymerized with UV light or simply by heating. The obtained thin films are composed of crystallites, all well oriented with respect to a given crystallographic direction of the substrate plane, as it was checked by X-rays and electron diffraction as well as by the nonlinear optical dichroism. The same substrate has been recently used for epitaxy of a noncentrosymmetric molecule: the p-methylbenzal-1,3 dimethyl-barbituric acid (MBDBA) (Le Moigne *et al.*, 1995). Preliminary SHG measurements show a sinificant second harmonic generation from epitaxied thin films (Le Moigne, *et al.*, 1995). Figure 3 shows second harmonic intensity as a function of incidence angle from a KAP substrate itself and KAP + epitaxied thin film. Although the KAP crystal has no center of symmetry and shows a significant SHG response, for this particular orientation no SHG from substrate itself was observed. Thus the observed SHG signal comes exclusively from the epitaxied thin film.

Several years ago Chollet *et al.* (1991) reported a significant second harmonic generation from sublimed copper phthalocyanine (CuPc) thin films. Anisotropic ESR experiments on thin films have shown that CuPc molecules are oriented, lying

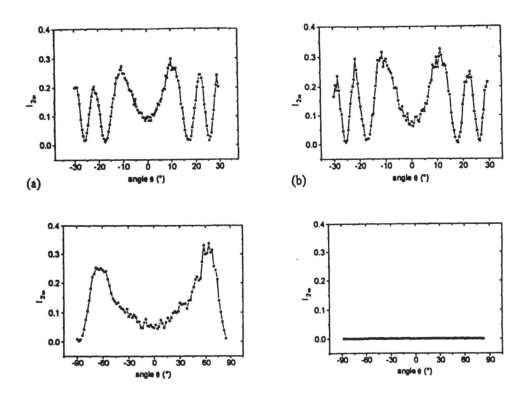

Figure 3 Incidence angle dependence of the intensity of the optical second harmonic generated by an epitaxied thin film on a substrate [figures on the left column labeled (a)] and by the substrate alone [figures on the right column labeled (b)]. The polarization configuration is *pp* for the two upper curves and *ss* for the two lower curves (after Le Moigne *et al.*, 1995).

almost flat with respect to the glass substrate, with a relatively large angular distribution (about 35 degrees) of molecular axis (perpendicular to the CuPc ring). In order to explain the origin of the SHG activity two possible explanation have been proposed:

(i) quadrupolar
(ii) resulting from the noncentrosymmetry of CuPc molecule at excited state. As a matter of fact, depending on the size of metal ion, this can be inside or outside of Pc ring. Copper ion radius is exactly at the limit. Thus it can be inside at fundamental state and outside at excited state. The recent, detailed measurements by Takezoe *et al.* (1996) show that the first hypothesis was correct. Note, that SHG has been also observed from fullerene C_{60} thin films (Hoshi *et al.*, 1991; Wang *et al.*, 1992; Kajzar *et al.*, 1993; Koopmans *et al.*, 1993). It is worthy to notice that for second order nonlinear optical activity no three dimensional order is required. A lack of center of symmetry in one direction is sufficient.

3.3 Static Field Poling

The dipolar molecules, like charge transfer molecules, can be oriented by applying static external field. As already mentioned, the orientation rate equation depends on the poling field and on the poling temperature, since thermal motion of molecules compete with dipolar interaction energy. In the case of polymeric systems the thin films are heated to the glass transition temperature, where chromophores become to be mobile and can rotate under the applied external field. Two static field poling techniques have been developed:

(i) contact (electrode) poling
(ii) corona poling.

In the first case one obtains in-plane poling if both electrodes are deposited on substrate. It is also possible to get perpendicular orientation when polymer is sandwiched between two electrodes. The main problem with this technique, limiting the poling efficiency, is dielectric breakdown, occuring at relatively low fields and which is due either to the "point effect", connected with charge injection or to the dielectric breakdown in surrounding air. Special care with electrode design is required to avoid this effect with a coating of poled film with a thick passive polymer layer or poling in high vacuum.

Corona poling (Comizzoli, 1987; Mortazavi *et al.*, 1989; Gadret *et al.*, 1991; Kajzar *et al.*, 1991) is done by applying a high voltage between a needle and a conducting bottom electrode (cf. Figure 4). Ionized charges are deposited on the thin film surface and create a high electric field inside, usually significantly higher than in electrode poling.

Both techniques are subject of charge injection into poled film, limiting poling fields and consequently the poling efficiency. Moreover, as already mentioned, polymers are heated to a high temperature. Such an elevated temperature may lead to thermally induced chemical reactions and consequently to the degradation of

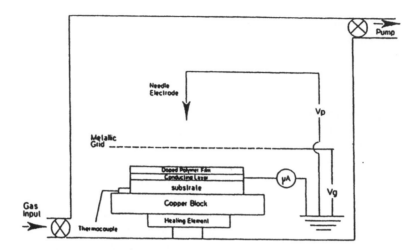

Figure 4 Schematic representation of a corona poling apparatus. Charges generated by an electric discharge on a metallic tip are deposited on the thin film surface and create a high electric field in the polymer film (after Gadret *et al.*, 1991).

material. The last two techniques photoassisted poling and all optical poling allow to obtain a polar orientation of molecules without heating ("cold poling").

3.4 Photoassisted Poling

Dumont and coworkers (Sekkat and Dumont, 1992; Dumont *et al.*, 1995) have observed that shining polymer thin films doped (or functionalized) with noncentrosymmetric dipolar azo dyes in the chromophore absorption band one observes a significant increase of electro-optic coefficient, corresponding to a better, polar orientation of chromophores. The measurements have been done using the attenuated total reflection technique and the optical field polarization was perpendicular to the applied low frequency external electric film to the thin film. A better stability of induced orientation was observed in the case of functionalized polymers than guest–host systems, as it is usually observed with static field poled polymers. The chromophores orient with dipolar moment perpendicular to the optical field (and parallel to the applied static (or low frequency) field. The chromophore orientation is going through trans-cis isomerization mechanism (cf. Figure. 5). The used azo dyes absorb strongly light if the exciting optical field is parallel to the dipolar transition moment. Excited molecules are mobile and change conformation to cis-form, with smaller volume. Subsequently they relax slowly to the trans form through nonradiative channels. A stable orientation is achieved when the dipolar transition moment is perpendicular to the optical, exciting field. Using a simple rate equation for the trans-cis izomerization process Dumont *et al.* (1995) describe well the observed temporal behaviour of electro-optic coefficient during attenuated total reflection measurements. While the rise on time is relatively fast (of the order of nanosecond, Kajzar *et al.* (1995), the recovery time is much slower.

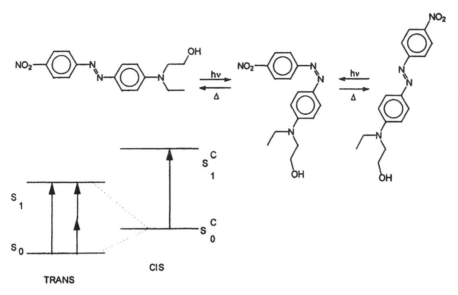

Figure 5 Orientation mechanism in photo-assisted and all-optical poling with azo-dye molecules. In photo-assisted poling trans-cis isomerization is achieved through one photon excitations, whereas in all optical poling through both one- and two-photon excitations (after Charra *et al.*, 1996).

3.5 All Optical Poling

Charra *et al.* (1993) have shown that one can obtain a polar orientation of chromophores in a functionalized or doped polymer film by purely optical fields. Initial experiments have been done in a nondegenerate four wave mixing geometry a permanent polarization with Disperse Red DR#1 chromophores, grafted on methyl methacrylate backbone. Two picosecond pump beams at 1.064 µm and a doubled frequency probe were used. The observed signal at 0.532 µm rose slowly, to the saturation value. Spontaneous SHG was observed after switching off the probe beam. The maximum second order nonlinear optical susceptibility value obtained in these experiments was of 3 pm/V (Charra *et al.*, 1993).

Significantly larger $\chi^{(2)}$ value was obtained in seeding geometry using two collinear picosecond beams at 1.064 µm and 0.532 µm. With the same polymer the best obtained value was of 76 pm/V, close to that obtained in corona poling (Nunzi *et al.*, 1995).

The mechanism of orientation is similar to that in photoassisted poling. The trans-cis izomerization is induced simultaneously by two photon transition and by harmonic photons (cf. Figure. 5). The "poling", nonzero temporal average field is obtained through the interference of input fields at ω and 2ω frequencies:

$$\langle\{E_{2\omega}\cos 2\omega t\,[E_\omega\cos(\omega t + \phi)]^2 + c\cdot c\}\rangle_t = E_{2\omega}E_\omega^2\cos(\phi), \tag{31}$$

where E's are amplitudes and ϕ's relative phases of corresponding fields.

Although the discussed here all optical poling has been demonstrated first with azo dyes, where its mechanism is quite clear, it has been also shown with other molecules, like octupolar molecules (Nunzi *et al.*, 1994a). It is important to note that up to now it is the only technique which allows to pole thin films with molecules which do not possess permanent dipoles. However the poling mechanism is not very clear.

4 THIN FILM LINEAR OPTICAL PROPERTIES CHARACTERIZATION TECHNIQUES

4.1 Refractive Index Determination

4.1.1 Refractive index and molecular orientation. Refractive indices are connected with the macroscopic linear polarizability tensor $\chi^{(1)}$. This tensor is at its turn connected with the molecular linear polarizabilities and orientations through Eq. (3). The organic films are made of optically active chromophores dissolved or chemically attached to a polymer matrix. Introducing chromophores results in an increase δn of the refractive index due to the larger optical polarizabilities of those chromophores. In addition, when the nonlinear molecules are poled in order to give to the film electro-optic properties, the film acquires an optical anisotropy with an extraordinary index n_e in the direction of poling and an ordinary one in the perpendicular plane. If we consider the chromophores as linear molecules, their orientation can be characterized by the angle Θ of their axis with the poling direction. It can be shown (Page *et al.*, 1990) that the order parameter S is related to the optical anisotropy by

$$S = \frac{3\langle\cos^2\Theta\rangle - 1}{2} = \frac{\delta n_e^2 - \delta n_o^2}{\delta n_e^2 + 2\delta n_o^2}. \tag{32}$$

Owing to the Clausius-Mosotti relation, $\delta n_e + 2\delta n_o^2$ is constant, depending only on the chromophore concentration. If the poling direction is perpendicular to the film plane, the absorption at normal incidence $A_\perp$ decreases and again the order parameter is connected to this variation (Page *et al.*, 1990):

$$\frac{\Delta A_\perp}{A_\perp} = S. \tag{33}$$

If poling is achieved by an external electric field **E** and under the assumption that chromophore interactions can be neglected, the order parameter can be related to the poling field through the following expression:

$$S = \frac{1}{15}\left(\frac{\mu E^*}{kT}\right)^2, \tag{34}$$

where μ is the chromophore permanent dipole and E^* is the local poling field connected with the applied external field by:

$$E^* = f^\circ \mathbf{E}, \tag{35}$$

where f° is the Onsager local field factor (cf. Eq. (25))

4.1.2 Common features in the different technique of refractive index determination. In the following we will restrict our discussion to the case when the films are uniaxial with their optical axis perpendicular to their planes, with ordinary index n_o and extraordinary n_e. At a given wavelength the refractive indices are deduced from the measurement of the transmission or reflection. Consequently we will briefly recall their theoretical expressions. At the interface of two semi-infinite media labeled i and $i + 1$ the amplitude of transmission of the field component in the interface plane (**E** for s-polarization and **H** for p-polarization) and reflection are given by:

– for the transverse electric polarization (labeled TE or s)

$$t_{i,i+1}^s = \frac{2k_{is}}{k_i^s + k_{i+1}^s}, \tag{36}$$

$$r_{i,i+1}^s = \frac{k_i^s - k_{i+1}^s}{k_i^s + k_{i+1}^s}, \tag{37}$$

where k_i^s is the z-component of the wave vector inside film i, which is given by

$$k_i^s = \frac{2\pi}{\lambda}(n_{o,i}^2 - n_a^2 \sin^2 \theta_a)^{1/2}. \tag{38}$$

In this last equation λ is the wavelength in vacuum, θ_a is the incidence angle in the ambiant (usually air) of refractive index n_a

– for the transverse magnetic polarization (labeled TM or p)

$$t_{i,i+1}^p = \frac{2k_i^p/n_{o,i}^2}{k_i^p/n_{o,i}^2 + k_{i+1}^p/n_{o,i}^2}, \tag{39}$$

$$r_{i,i+1}^p = \frac{k_i^p/n_{o,i}^2 - k_{i+1}^p/n_{o,i}^2}{k_i^p/n_{o,i}^2 + k_{i+1}^p/n_{o,i}^2} \tag{40}$$

with

$$k_i^p = \frac{2\pi}{\lambda}\left(n_{o,i}^2 - \frac{n_{o,i}^2}{n_{e,i}^2}n_a^2 \sin^2 \theta_a\right)^{1/2}. \tag{41}$$

In the case of a 3-medium layer (film – labeled 2 – of thickness h_2 sandwiched between two semi-infinite media 1 and 3), the reflection coefficient r_{13} inside medium 1 and transmission coefficient t_{13} from medium 1 to medium 3 are given by the Airy formulae:

$$t_{13} = \frac{t_{12}t_{23}e^{i\varphi_2}}{1 + r_{12}r_{23}e^{2i\varphi_2}}, \tag{42}$$

$$r_{13} = \frac{r_{12} + r_{23}e^{2i\varphi_2}}{1 + r_{12}r_{23}e^{2i\varphi_2}}, \tag{43}$$

where φ_2 is the phase inside the film given by k_2h_2.

In the case of a stratified structure, an iteration has to be performed on the Airy formulae starting from the last semi infinite medium encountered by the monochromatic beam.

4.1.3 Transmission at normal incidence.

This method consists of measuring at normal incidence the transmission of a film deposited on a substrate, as a function of the wavelength. In this simple case the transmission and reflection at the interface $i, i+1$ are given by:

$$t_{i,i+1} = \frac{2n_{o,i}}{n_{o,i} + n_{o,i+1}}, \tag{44}$$

$$r_{i,i+1} = \frac{n_{o,i} - n_{o,i} + 1}{n_{o,i} + n_{o,i+1}}, \tag{45}$$

because of the phase term in the denominator of t_{13} (cf. Eq. (42)), the transmission reaches extrema. Taking account of the incoherent multiple reflection inside the substrate the values of these maxima are:

$$T = \frac{(t_{12}t_{23}t_{31})^2}{(1 \pm r_{12}r_{23})^2(1 - r_{31}^2 r_{32}^2)}, \tag{46}$$

where t_{31} is the transmission coefficient from the substrate 3 to the ambiant 1 $2n_3/(n_3 + n_1)$; r_{31} and r_{32} are the Fresnel coefficients inside the substrate at the interfaces with the ambiant and the film respectively. Rigorously r_{32} should be calculated by taking account of the Airy formula, but in practice it does not affect the results to use the simple Fresnel coefficients. The nature of the extrema (maxima or minima) in relation with the sign $\pm$ depends on the sign of r_{23} which is connected with the relative magnitude of the film and substrate indexes. The comparison of the experimental value with the theoretical one given by Eq. (46) enables to determine the refractive index at the extrema of the transmission. Explicit expressions of those extrema can be found elsewhere (Heavens, 1964).

4.1.4 Abeles method.

This method (Abeles, 1950) is based on the fact that for the Brewster angle of incidence θ_{12}^B at the interface ambient/film (in TM polarization of course) the reflection is the same as it would be on the bare substrate. This can be easily verified from Eq. (43). For this angle r_{12} is equal to 0 which yields $r_{13} = r_{23}e^{2i\varphi}$. But for this incidence angle the reflection amplitude on the bare substrate is also equal to r_{23}. For an isotropic film this gives the film refractive index directly:

$$n_2 = n_1 \tan\theta_{12}^B. \tag{47}$$

The experiment is performed by depositing the film on half of a substrate. This substrate is placed on a holder which enables a rotation with the axis perpendicular to its plane. The reflection measurement is performed successively on the bare and film covered substrate. If the detector is connected to a lock-in amplifier the incidence angle giving a zero signal can be determined very accurately leading to a determination of the refractive index within 10^{-3}

Morever, one can easily check that the above property holds for the Brewster angle at the interface film/substrate. Unfortunately for most of the substrates whose index is larger than 1.3 this angle corresponds to $n \sin\theta > 1$.

4.1.5   Ellipsometry. This common method (Azzam, 1977) consists of determining the ratio of the reflection amplitudes in the p and s polarizations. This ratio is written:

$$\frac{r_p}{r_s} = \tan\Psi e^{i\Delta}. \tag{48}$$

The general principles of determining Ψ and Δ is as follows: the sample is illuminated by a monochromatic collimated beam whose polarization is continuously varied (by means of a rotating polarizer). The reflected intensity is analyzed by a fixed polarizer, whose polarization direction can be however adjusted. The amplitude and phase (with respect to the rotation of the first polarizer) of the intensity reaching the detector is analyzed by a lock-in amplifier triggered by the rotating polarizer. Since two parameters are determined, an experiment performed at a fixed incidence angle allows the determination the real and imaginary parts of an isotropic semi-infinite medium. For a transparent isotropic film deposited on a substrate of known index the same experiment gives its thickness and real refractive index. but if the film is optically anisotropic and in addition absorbing, the incidence must be varied to determine the 5 parameters (h, $n_o + i\kappa_o$, $n_e + i\kappa_e$). The relative accuracy obtained by ellipsometry is within 10^{-3}.

4.1.6   Mode coupling (m-lines). This method, initiated at the end of the sixties (Tien, 1969) and adapted to anisotropic films, consists of exciting the guided mode in a film of thickness h via a high refractive index prism (Figure 6). A guided mode is a wave, stationary in the perpendicular direction to the film, which propagates in the film with a wavevector $k_x = \beta k_v$. (β, ratio with the wavevector k_v in vacuum is the effective index of the mode). The stationary condition results in discrete values of the transverse wave component k_z (Tien 1969):

$$2k_{z,\text{pol}}^{(m)}h + 2\Psi_{\text{pol}}^{(m)} + 2\Phi_{\text{pol}}^{(m)} = 2m\pi, \tag{49}$$

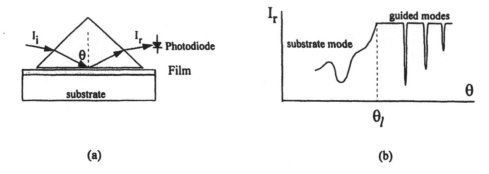

(a)   (b)

Figure 6 Attenuated total reflection (ATR) experimental set up for measuring the effective index of the modes in a film (a). Reflected intensity variation with the incidence angle (b). For angles smaller than the angle ql the modes are no more guided but leaking in the substrate resulting in a much larger FWHM.

m (order of the mode) is an integer ≥ 0 which is equal to the node number inside the guide; pol is the polarisation (s or p). Ψ and Φ are the phase shift of the total reflections (given by the phase of the Fresnel coefficients; cf Eqs. (37) and (40)) at inner interfaces of the film with medium (air or substrate) of refractive index n:

$$\Psi_{TE} = -\tan^{-1}\left(\frac{\beta^2 - n^2}{n_0^2 - \beta^2}\right)^{1/2}, \tag{50}$$

$$\Psi_{TM} = -\tan^{-1}\left(\frac{n_0 n_e (\beta^2 - n_1^2)^{1/2}}{n^2 (n_e^2 - \beta^2)^{1/2}}\right). \tag{51}$$

The experimental determination of $\beta(k_x)$ allows the determination of k_z via the relations:

$$\frac{k_{x,s}^2 + k_{z,s}^2}{n_0^2} = \left(\frac{2\pi}{\lambda}\right)^2, \tag{52}$$

$$\frac{k_{x,p}^2}{n_e^2} + \frac{k_{z,p}^2}{n_0^2} = \left(\frac{2\pi}{\lambda}\right)^2. \tag{53}$$

When the incidence of the laser beam varies and the x-component of the wave-vector inside the prism matches that of a guided mode, the light intensity is launched into the film resulting in a dip of the reflection detected by the the diode. From the above set of equations, it is possible to determine the ordinary and extraordinary refractive indices within $(2 \div 3) \times 10^{-4}$. By fitting the shape of the reflection deep in the case when the coupling is perfectly known (for instance a thin metallic thin film deposited between the prism and the film) it is possible to determine the effective imaginary part of the refractive index (which originates from all kind of losses: intrinsic, domains boundary, surface roughness, etc.).

4.2 Propagation Losses Measurements

All components designed for applications especially in waveguiding configurations should have the lowest losses possible. One of the main reason is that it is very likely several such components are to be found successively which increases the losses. No practical devices can be used if its losses exceed 5 dB

There are three main kinds of losses: insertion, propagation and radiation. Insertion losses are connected to the optical quality of the entrance face surface of the guide and to the overlap integral between the guided mode and the laser beam. Propagation losses are of two kinds: the first category includes intrinsic losses due to the absorption of the molecules. In the telecommunication wavelengths (1.3 and 1.55 μm) the combination bands of the CH and OH vibrations for polymer guides generate losses ($\sim$ 1 dB/cm: Norwood, 1992) which can be lowered by substitution of the atom H by a deuterium or a halogen atom. The second category arises from all kinds of inhomogenities which result in a spatial refraction index variation. Such kind of losses generates Rayleigh scattering.

The radiation losses are essentially due to the shape of the guide (curvature radius as compared to the transverse width, angle between embranchments) and can be avoided by an electromagnetic modelization such as "beam propagation methods". Thickness inhomogenities, not always predictable, may also induce losses (Hunsperger, 1982).

Loss measurement techniques

The aim of most of the studies is to determine propagation losses which are connected with the variation of guided intensity with propagation length.

(1) The first technique consists of measuring the intensity exiting from guides of different lengths. Either several guides of different lenghts are fabricated for that purpose, or a preexisting one is cut in order to reduce its length. The main drawbacks of this method is the reproducibility of the input coupling and the guide fabrication. In addition the cutting method is destructive.

(2) The problem of insertion is overcome by keeping the same insertion device and sliding a decoupling prism (which theoretically extracts all the guided light) along the guide. In order to avoid perturbation of the input coupling by strains, an index matching fluid is inserted between the decoupling prism and the film (Zernike, 1971). More sophisticated technique using prisms have enabled to measure losses as low as 0.02 dB/cm (Weber, 1973).

(3) Propagation losses can be deduced from the amount of scattered light along the guide, if the guide is assumed to be homogenous. This scattered light is collected by an optical fiber moving along the guide (Osterberg, 1964), technique which was used for Langmuir–Blodgett films (Küpfer, *et al.*, 1992). The scattered light can be also directly imaged by a lens on a photodiode array (Franke, 1986) and losses in polymeric films were measured using this technique (Chollet *et al.*, 1992).

The three techniques described previously, which enable to measure propagation losses, do not discriminate between scattering and absorption ones. Photothermal deflection (Boccara, 1989) does discriminate and has been applied to organic films (Skumanich, 1993). The guiding film is immersed in a nonsolvent liquid and a cw laser bearm propagates in the liquid, parallel to the film surface and very close to it. Then the film is illuminated transversally by an intense beam at the wavelength one wants to measure its absorption. If the films absorbs, it results in a temperature rise in the liquid and subsequently a deflection of the first laser beam (mirage effect). On the opposite, scattering does not correspond to energy absorption and consequently does not induce temperature change in the liquid.

5 SECOND HARMONIC GENERATION TECHNIQUES

Several techniques have been developed for thin film second order nonlinear optical properties characterization. The most commonly used is transverse second harmonic generation method. The principal advantage of this technique is the measure of the electronic response of material. As a matter of fact the response time in this technique is of the order of 10^{-15} s, as the harmonic wave must follow the fundamental wave. This technique requires a change of the propagation length in material which is usually achieved by rotating thin film around an axis perpendicular to the beam propagation direction. The SHG measurements can be also performed *in situ* allowing a real time study of the cinetics of polarization and relaxation processes (Dantas-Demorais *et al.*, 1996). Usually thin films are deposited on one side of substrate only. In that case the harmonic intensity is given by

$$I_{2\omega}^{ffh} = \frac{\pi^2}{2\varepsilon_0} \left| \frac{\chi_{ffh}^{(2)}(-2\omega;\omega,\omega)}{\Delta\varepsilon} \right|^2 |P_{fh}(\theta)|^2 |T_{fh}(\theta)(e^{i\Delta\varphi} - 1)|^2 (I_\omega)^2, \tag{54}$$

where T_{fh}s are transmission and boundary condition factors for a given fundamental (f) and harmonic (h) polarization configuration. Similarly P_{fh}, is projection factors and $\Delta\varphi$ is the phase mismatch between fundamental and harmonic beams (for details see Swalen, 1996).

In same cases, like in Langmuir-Blodgett technique, thin film is deposited on both sides of a centrosymmetric substrate. In that case harmonic fields generated in both films interfere and the result of interference depends on refractive index dispersion in substrate itself The output SHG intensity is then given by

$$I_{2\omega}^{ffh} = \frac{\pi^2}{2\varepsilon_0} \left| \frac{\chi_{ffh}^{(2)}(-2\omega;\omega,\omega)}{\Delta\varepsilon} \right|^2 |P_{fh}(\theta)|^2$$
$$\times \left| T_{fh}^1(\theta)(e^{i\Delta\varphi_1} - 1) + T_{fh}^2(\theta)(e^{i\Delta\varphi_2} - 1)e^{i(\Delta\varphi_{1,2}^r + \Delta\varphi_{1,2}^i)} \right|^2 (I_\omega)^2 \tag{55}$$

where again $\Delta\varphi$'s are the phase mismatches in different media and T's are the transmission factors. Eqs. (54)–(55) take account of thin film absorption by introducing complex refractive index $n = n^r + i\kappa$ where κ is the extinction coefficient (Swalen, 1996).

Usually SHG intensities are calibrated with second harmonic generation using a quartz single crystal ($d_{11} = \chi^{(2)}_{111}(-2\omega; \omega, \omega)/2 = 0.48$ pm/V (Choy *et al.*, 1976).

For a very crude estimation, assuming refractive indices of polymer thin films close to those of reference quartz, one uses a simplified formula, which for an s–p fundamental-harmonic beam configuration reads

$$\chi^{(2)}_{sp} = \frac{2}{\pi} \chi^{(2)}_Q \left[\frac{I^{2\omega}_P}{I^{2\omega}_Q}\right]^{1/2} \frac{I^Q_c}{l} \frac{1}{\cos\theta^{max}_\omega \sin\theta^{max}_{2\omega}} \tag{56}$$

and

$$\chi^{(2)}_{pp} = \frac{2}{\pi} \chi^{(2)}_Q \left[\frac{I^{2\omega}_P}{I^{2\omega}_Q}\right]^{1/2} \frac{I^Q_c}{l} \frac{1}{\cos\theta^{max}_\omega \{\sin\theta^{max}_{2\omega}[1+(a-1)\cos^2\theta^{max}] + a\sin2\theta^{max}\cos\theta^{max}\}}, \tag{57}$$

where subscripts (or superscripts) P, Q refer to the polymer film and reference quartz respectively, $I^{2\omega}$ is the maximum intensity (at normal incidence in the case of quartz, measured in the p–p configuration with E parallel to the x-axis, θ^{max} is angle at which the maximum of harmonic intensity from thin film is measured, $I^Q_c = \lambda/4(n_{2\omega} - n_\omega)$ is the coherence length of quartz l is the thin film thickness and $a = d_{13}/d_{33}$. It is evident that in order to get the diagonal tensor component d_{33} value it is necessary to make SHG measurements in s–p configuration, leading to the determination of d_{13} which, injected into Eq. (56) allows its determination.

6 ELECTRO-OPTIC EFFECTS

6.1 Generalities

6.1.1 Phenomenological approach. As stated before, the electro-optic phenomena are the refractive index variations induced by an applied electric field whose frequency is several orders of magnitude below that of the optical wavelength. In general and specially in electro-optic organic films, the refractive index is anisotropic (birefringence). We will write the equations in the reference frame where it is diagonal. The expansion of the refractive index variation, up to the second order in the applied field **E** is

$$\delta n_{ii} = p_{ij}E_j + k_{ijk}E_jE_k, \tag{58}$$

p_{ij} are the Pockels tensor components which correspond to the linear electro-optic effect and k_{ijk} are the Kerr tensor components which corresponds to the quadratic electro-optic effect.

Let us at this point enter into very simple but very important symmetry considerations: if the material we are dealing with (either crystalline or amorphous) is centrosymmetrical, its two electro-optical tensors are invariant by reversing the direction of the frame axes, neither δn_{ii} which is a scalar. On the other hand, in such a transformation, the electric field components E_j change of sign and Eq. (58) can be fulfilled only if the tensor p is identically equal to zero, which is not the case for k

whose magnitude can have any value, irrespectively of the material symmetry. This symmetry restriction (nonlinear electro-optic effects in centrosymmetrical compounds) is of prime importance for applications.

6.1.2 Connection with nonlinear optical polarizabilities.

In the case when simultaneously a material is submitted to an electromagnetic wave $\mathbf{E}_\omega \cos \omega t$ and a dc field $\mathbf{E}_0$ the polarizability variation $\Delta \mathbf{P}_\omega$ (at ω frequency) depends on the dc-field according to

$$\Delta P^i = \varepsilon_0 \left(\chi_{ijk}^{(2)} (E_0^j + E_\omega^k \cos \omega t)^2 + \chi_{ijkl}^{(3)} (E_0^j + E_0^k + E_\omega^l \cos \omega t)^3 \right). \tag{59}$$

Consequently the variation of polarization at the optical frequency ω is written:

$$\Delta P_\omega^i \cos \omega t = \varepsilon_0 \left(2\chi_{ijk}^{(2)} E_0^j + 3\chi_{ijkl}^{(3)} (E_0^j)^2 + 6\chi_{ijkl}^{(3)} E_0^j E_0^l \right) E_\omega^k \cos \omega t. \tag{60}$$

In Eq. (60) the term in the bracket is the variation of the optical dielectric constant $\varepsilon_{jk}(\omega)$. Working in the diagonal frame, we have $\Delta n_{ii}/\Delta n_{ii} = \Delta \varepsilon_{ii}/2\varepsilon_{ii}$ which yields in combination with (60):

$$\Delta n_{ii} = \frac{1}{n_{ii}} \left[\chi_{iij}^{(2)} E_0^j + \frac{3}{2} \chi_{iijj}^{(3)} (E_0^j)^2 + 3\chi_{iijk}^{(3)} E_0^j E_0^k \right]. \tag{61}$$

In almost all experimental configurations the electric field is applied along the principal axes of the index ellipsoid with the consequence that the last term in Eq. (61) drops on. The comparison of Eqs. (58) and (61) enables to calculate the value of the Pockels and Kerr tensors. However, for historical reason, the electro-optical tensor relates the reciprocal of the optical dielectric tensor to the electric field. This originates because when these effects were discovered at the end of the last century, the equation of the ellipsoid index was written (in its principal frame): $\sum_{i=1}^{3} A_i x_i^2 = 1$ the variation of A_i was expanded into the power of the field. Since $A_i = (\varepsilon_{ii})^{-1} = (n_{ii})^{-2}$ the expansion of the refractive index is written as follows:

$$\Delta n_{ii} = \frac{n_{ij}^3}{2} [r_{ii} E_0^j + s_{ij} (E_0^j)^2]. \tag{62}$$

The comparison between (61) and (62) yields the classical relations:

$$r_{ij} = \frac{2\chi_{iij}^{(2)}}{n_{ii}^4}, \qquad s_{ij} = \frac{3\chi_{iijj}^{(3)}}{n_{ii}^4}. \tag{63}$$

The general principles of all the experimental setup is to connect the modulation of the optical response (transmittance or reflection) with the physical parameter (mainly refractive index but also film thickness as explained below) modulations. In order to apply the field, the film is sandwiched between two electrodes; one at least must be transparent in order to allow the probe monochromatic beam to travel through the film. Since the transmittance or reflection variations are always very small $(10^{-5}-10^{-4})$ the detection is performed with a lock-in amplifier tuned either to the applied field frequency Ω $(0.5-10\,\text{kHz})$ or to its second harmonic. From Eq. (62) it yields that the amplitudes of the refractive index modulations at the fundamental frequency Ω and at the second harmonic 2Ω are equal to:

$$\Delta n_{ij}^{\Omega} = \frac{n_{ii}^3}{2} r_{ij} E_j^{\Omega}, \qquad \Delta n_{ij}^{2\Omega} = \frac{n_{ii}^3}{2} s_{ij} (E_j^{\Omega})^2. \tag{64}$$

In practice the experimental interpretation is complicated by the fact that the films undergo a thickness modulation upon the application of the electric field (piezo and electrostriction effects).

6.1.3 Electro-mechanical effects

6.1.3.1 Thickness modulation induced by the electric field

As for the refractive index, the application of an electric field induced strengths to the materials which result in a contraction or expansion. In the case of a thin film which is strongly clamped to its substrate, this contraction or expansion concerns only the thickness h of the film (in the direction labeled 3, perpendicular to the film plane). As for the refractive index, the thickness variation can be expanded as a power of the electric field:

$$\frac{\Delta h}{h} = d_{33}E - c_{333}E^2. \tag{65}$$

The linear term in E corresponds to the piezo-electric effect and the quadratic term to the electro-striction effect (including the attraction of the electrodes if they are directly in contact with the film). Since the thickness is a scalar, the same symmetry considerations as for the electro-optic effect hold for the electro-mechanical one: the piezo-electric effect is present only in noncentrosymmetrical materials, while the electro–striction (including the attraction of the electrodes) is always present. In addition it should be reminded that the electro-striction is always a contraction of the materials. This is why we have written a minus sign before c_{333} which is a positive quantity. On the opposite hand the piezo-electric effect leads to either an expansion or a contraction, depending on the relative orientation between the electric field and the molecules inside the film. As we have previously stated, the sample transmission (or reflection) modulation is dependent on the optical path modulation which is connected with both the refractive index and the thickness. Moreover, as we will show now, an indirect refractive index modulation originates from the thickness modulation.

6.1.3.2 Indirect refractive index modulation due to the electro-mechanical effects

In the introduction we have shown that in molecular crystal there is a connection between the molecular optical susceptibility and the macroscopic one (gas phase model) and that the macroscopic susceptibilities are is linearly dependent on the molecule volumic concentration N. Consequently an increase in N (decrease of the thickness h) results in an increase in the refractive index.

For the quantitative formulation, we will restrict our model to the case of a film made of axial molecules. These molecules have a polarizability $\alpha_\parallel$ along their axes $\mathbf{a}$ and $\alpha_\perp$ in the perpendicular directions. Consequently their linear susceptibility tensor has the following expression:

$$\vec{\alpha} = \alpha_\parallel \|\mathbf{a}\rangle\langle\mathbf{a}| + \alpha_\perp (\vec{1} - |\mathbf{a}\rangle\langle\mathbf{a}|). \tag{66}$$

The molecule axes have a polar orientation with respect to the normal to the plane $\mathbf{z}(\angle \mathbf{z}, \mathbf{a} = \Theta)$. These hypotheses describe the most encountered situations and in particular the case of poled polymer films.

The linear macroscopic susceptibility tensor $\chi^{(1)}$ can be expressed in two ways: First it is directly connected to the refractive index:

$$n_{ii}^2 - 1 = \chi_{ii}^{(1)}. \tag{67}$$

Secondly the gas phase model connects it to the molecular susceptibility:

$$\chi_{ii}^{(1)} = \frac{n_{ii}^2 + 2}{3\varepsilon_0} N\alpha_{\text{eff}}, \tag{68}$$

where α_{eff} takes into account the projection factors of the tensor components

$$\alpha^{\text{eff}} = \alpha_\perp + \frac{\alpha_\parallel - \alpha_\perp}{2} \langle \sin^2\Theta \rangle \quad \text{for } i = 1, 2 \text{ (ordinary index)} \tag{69}$$

and

$$\alpha^{\text{eff}} = \alpha_\parallel \langle \cos^2\Theta \rangle + \alpha_\perp \langle \sin^2\Theta \rangle \quad \text{for } i = 3 \ (z) \text{ (extraordinary index).} \tag{70}$$

Equating (67) and (68) one can easily deduce the variation Δn of the refractive index:

$$2n_i \Delta n_i = \frac{(n_i^2 + 2)(n_i^2 - 1)}{3}\left(\frac{\Delta\alpha_i^{\text{eff}}}{\alpha_i^{\text{eff}}} + \frac{\Delta N}{N}\right). \tag{71}$$

This last equation relies the refractive index variation to the various phenomena induced by the applied field. In the majority of the case the dominant term is the modulation of $\alpha_\parallel$ and $\alpha_\perp$ by the electric field (Pockels or Kerr effects). In piezo-electric films, the thickness modulation Δh (connected to the molecule density modulation ΔN by $\Delta h/h = -\Delta N/N$ since the film is clamped on the substrate) can bring an important contribution. In addition to the electro-optic and electro-mechanic effects, a modulation of the refractive index is generated by the oscillations induced by the applied field E^Ω. This results in a modulation of Θ which induces a modulation of α^{eff} (cf Eqs. (68) and (70) and consequently of the refractive index (cf Eq. (68)). The molecules are coupled to the electric field through their large permanent dipole μ

(Kuzyk *et al.*, 1990a) and the oscillation can be related with the free volume around the chromophore (Boyd *et al.*, 1991). This orientational effect can contribute substantially to the dc-Kerr effect (Kuzyk *et al.*, 1990b). The frequency dependence of this molecule oscillation has been also studied recently by second harmonic generation (Sugihara *et al.*, 1996).

6.2 Electro-optic Coefficient Determination Techniques

A thorough review of the different techniques used for determing the electro-optic coefficients has been published recently (Chollet *et al.*, 1996). Before briefly reviewing these techniques we would like to describe their general common features.

6.2.1 General features. All the experiments used for the electro-optic coefficient determination are based on the variation of the optical response $\Re$ (transmission or reflection) of the sample when it is submitted to an external applied field $E\cos\omega t$. This response $\Re((p_1,p_2,\ldots p_m)$ depends on m parameters p_i and can be expanded to the second order of the applied field E:

$$\Re(E) = \sum_i \frac{\partial \Re}{\partial p_i}\left(\frac{\partial p_i}{\partial E}E + \frac{1}{2}\frac{\partial^2 p_i}{\partial E^2}E^2\right) + \frac{1}{2}\sum_i \frac{\partial^2 \Re}{\partial p_i^2}\left(\frac{\partial p_i}{\partial E}E\right)^2. \tag{72}$$

The derivatives $\partial\Re/\partial E$ and $\partial^2\Re/\partial p_i^2$ can be calculated according to the sample configuration. The aim of the studies is to determine the derivatives $\partial p_i/\partial E$ and $\partial^2 p_i/\partial E^2$ which are respectively the Pockels and Kerr electro-optical coefficients (when p_i are the refractive indices) and the piezo and electro-striction coefficients (when p_i is the thickness).

6.2.1.1 Measurement with a sinusoidal applied field
As already pointed out, the time variation of E is almost always sinusoidal of frequency Ω and the detection of $\Re$ is performed at Ω and at 2Ω using a lock-in amplifier. It is straightforward from (72) that the responses at Ω and 2Ω are:

$$\Re_\Omega(E) = \sum_i \frac{\partial \Re}{\partial p_i}\frac{\partial p_i}{\partial E}E, \tag{73}$$

$$\Re_{2\Omega}(E) = \frac{1}{2}\sum_i\left[\frac{\partial \Re}{\partial p_i}\frac{\partial^2 p_i}{\partial E^2} + \frac{\partial^2 \Re}{\partial p_i^2}\left(\frac{\partial p_i}{\partial E}\right)^2\right]E^2. \tag{74}$$

In (74) the second term in the bracket is the cascading effect where $\chi^{(3)}$ has the same effect as $\chi^{(3)}$ through the nonlinearity of the process.

In many papers the terms involving $\partial^2\Re/\partial p_i^2$ are neglected without justification and consequently the modulation at 2Ω is only attributed to $\partial^2 p_i/\partial E^2$ which corresponds to the quadratic effects. Fortunately in most (but not all) cases the experimental parameters are choosen in order to optimize the response at Ω which means that the derivatives $\partial\Re/\partial p_i$ reach their maximum values and consequently that $\partial^2\Re/\partial p_i^2$ are effectively close to zero. In addition we will show that even if

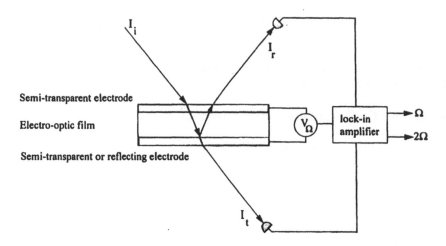

Figure 7   General experimental setup for measuring the electro-optic coefficients of thin films.

$\partial^2 \Re / \partial p_i^2$ is not equal to zero the second term in (74) is generally negligible besides for film thinner than the wavelength. If the parameter p_i is the refractive index n and h the thickness of the sample, the order of magnitude of $\partial \Re / \partial n$ and $\partial \Re / \partial n^2$ are respectively $(2\pi h / \lambda)\Re$ and $(2\pi h / \lambda)^2 \Re$. From Eq. (69) it appears that the cascading term is $\sim (1/2n)(2\pi h/\lambda)^2 \Re(\chi^{(2)}E/n)^2$ while the one involving $\chi^{(3)}$ is $\sim (3/2n)(2\pi h/\lambda)$ $\Re(\chi^{(3)}E^2/n)$. The ratio of these two terms is $(2\pi h/3\lambda)\,(\chi^{(2)})^2/\chi^{(3)}$.

Typical values of the parameters are $n = 1.6$; $r = 20\,\text{pm/V}$ $(\chi^{(2)} = 60\,\text{pm/V})$ and $\chi^{(3)} = 5 \times 10^{-20}\,\text{m}^2/\text{V}^2$ yields to a ratio equal to 0.4 h/λ. This shows that for film thickness smaller than the wavelength the cascading effect is negligible. In any case the magnitude of $\partial^2 \Re / \partial n^2$ can be calculated either explicitly or numerically.

6.2.1.2   Measurement with the superimposition of a dc-field to the sinusoidal one
In order to measure simultaneously at the fundamental frequency the Pockels and Kerr the applied field was the sum of a dc-field and an ac-field (Röhl *et al.* 1991).

$$\mathbf{E}(t) = \mathbf{E}_{dc} + \mathbf{E}_{ac} \cos \Omega t. \tag{75}$$

It yields from (73) that the modulated response at the frequency Ω is

$$\Re_\Omega(E_{ac}, E_{dc}) = \left\{ \frac{\partial \Re}{\partial p_i} \left(\frac{\partial p_i}{\partial E} + \frac{\partial^2 p_i}{\partial E^2} E_{dc} \right) + \frac{\partial^2 \Re}{\partial p_i^2} \left(\frac{\partial p_i}{\partial E} \right)^2 E_{dc} \right\} E_{ac} \cos \Omega t. \tag{76}$$

Of course as previously there is always a cascading effect (last term in the bracket). For the refractive index we have

$$\frac{\partial n_i}{\partial E_k} = \frac{\chi_{iik}^{(2)}}{n_i}, \tag{77}$$

$$\frac{\partial^2 n_i}{\partial E_k^2} = \frac{3\chi_{iikk}^{(3)}}{n_i}. \tag{78}$$

From Eqs. (76) and (77) it appears that $\chi^{(2)}$ is directly obtained when there is no dc-field and the variation with the dc field gives exactly the same combination of $\chi^{(3)}$ and $\{\chi^{(2)}\}^2$ as the modulation at $2\Omega : \Re_{2\Omega}$.

6.2.2 Experimental techniques

6.2.2.1 Mach–Zehnder interferometer
The optical scheme of this classical technique is represented in Figure 8. A wedge slab in the reference arm enables to adjust the phase mismatch Φ between the two arms (reference and sample). After crossing the sample (thickness h and ordinary index n_o), the wave phase in the reference arm is

$$\delta = \frac{2\pi n_o h}{\lambda}. \tag{79}$$

If the intensity in those arms are I_1 and I_2 respectively, the total intensity after the waves recombine is

$$\Re = \tfrac{1}{2}\{I_1 + I_2 + 2\sqrt{I_1 I_2}\cos(\Phi - \delta)\}. \tag{80}$$

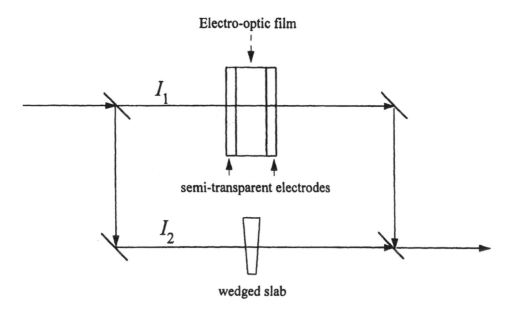

Figure 8 Mach–Zehnder interferometer for measuring the electro-optic coefficients of thin films and crystals.

The combination of (79) and (80) yields:

$$\frac{\partial \mathfrak{R}}{\partial n_o} = \frac{2\pi h}{\lambda} \sqrt{I_1 I_2} \sin(\Phi - \delta), \tag{81}$$

$$\frac{\partial^2 \mathfrak{R}}{\partial n_o^2} = -\left(\frac{2\pi h}{\lambda}\right)^2 \sqrt{I_1 I_2} \cos(\Phi - \delta). \tag{82}$$

When the sample is submitted to a transverse voltage $V \cos \Omega t$ (along the axis 3) due to the Pockels effect, the ordinary index is modulated according to

$$n_o(t) = n_o + \frac{r_{13} n_o^3}{2h} V \cos \Omega t \tag{83}$$

which gives the response at the modulation frequency Ω

$$\mathfrak{R}_\Omega = \frac{\pi r_{13} n_o^3 V}{\lambda} \sin(\Phi - \delta) \sqrt{I_1 I_2} \tag{84}$$

This technique is easy to setup, but since the interference is made at the laboratory scale, requires a monochromatic source of high spatial coherence. It is also very sensitive to vibrations. It has been used for the determination of Pockels coefficients in crystals (Hierle, 1980).

By tuning the detection at 2Ω, it is in principle possible to determine the dc-Kerr effect. However Eq. (82) shows very clearly the increasing contribution of the cascading effect with the sample thickness. In the present case, the ratio between $\partial^2 \mathfrak{R}/\partial n^2$ and $\partial \mathfrak{R}/\partial n$ is exactly $2\pi h/\lambda$ when $\Phi - \delta = \pi/4$ and all the considerations of Section 6.2.1 apply. The balance between the two derivatives can be varied by ajusting Φ with the wedged slab. Since it is difficult to set it directly at a position which corresponds to a zero of $\partial^2 \mathfrak{R}/\partial n^2$ (by maximizing the $\mathfrak{R}_\Omega$ response which corresponds to a maximum of $\partial \mathfrak{R}/\partial n$) it is much preferable to determine two consecutive zeros of $\mathfrak{R}_\Omega$ and setting the slab exactly at mid-position. In this configuration $\chi^{(3)}$ is directly linked to $\mathfrak{R}_{2\Omega}$ by

$$\mathfrak{R}_{2\Omega} = \frac{3\pi \chi_{1133}^{(3)}(-\omega; \omega, 0, 0) h}{\lambda} E_\Omega^2. \tag{85}$$

6.2.2.2 Fabry–Perot method

This method consists in studying the transmission variation of a Fabry–Perot etalon connected with the refractive index variation. The value of this transmission is connected to the transmission, reflection and absorption coefficients of the metallic layers:

$$T = \frac{T_m^2}{(T_m + A_m)^2} \frac{1}{1 + F \sin^2 \varphi}, \tag{86}$$

where $F = 4R_m/(1 - R_m)^2$ is the cavity resonance parameter and $\varphi = k_z h + \varphi_m$ the total phase-shift including the contribution of the metallic reflections φ_m. The variation of the transmission induced by the electric field is given by

$$\Delta T = \frac{\partial T}{\partial \varphi} \Delta \varphi = \varphi \frac{\partial T}{\partial \varphi} \left(\frac{\Delta n}{n} + \frac{\Delta h}{h} \right), \tag{87}$$

Δh is given by Eq. (62) and Δn by Eqs. (59), (65) and (71). Selective detection of ΔT^{Ω} and $\Delta T^{2\Omega}$ enables to determine the linear and quadratic response of ΔT, but only if the cascading effects can be neglected.

By varying the incidence angle it was possible to determine the Pockels and Kerr effects and the amount of charges trapped during the poling process (Ederling, 1989). Pockels and Kerr coefficients were determined by studying at varible wavelength the electric field induced modulation in a Fabry–Perot interferometer (Uchiki, 1988; Kobayashi, 1989a; 1989b). These studies were performed in polycarbonate films doped with oriented diethyl-amino-nitrostilbene molecules (DEANS).

6.2.2.3   Attenuated total reflexion

This technique consists of measuring the modulation of the m-lines induced by the applied field. When we have described the refractive index measurement technique, we have seen that the knowledge of the propagation constants (effective index) of two guided modes in the s-polarization (TE) and at least one in the p-polarization (TM) enabled to determine very precisely the two refractive indices n_o and n_e and the thickness h. Performing the same experiment with an applied field gives the variations of those three parameters and consequently the electro-optic and electro-mechanic coefficients (Dentan, 1989; Morichère 1989; Dumont, 1991). Owing to the small variation of the refractive indices, the time variation of the field is sinusoidal at the audio-frequency Ω and the time dependence of the modulated reflected beam is analyzed by a lock-in amplifier. For films whose thicknesses is smaller or of the same magnitude than the wavelength, cascading effects may be neglected. In that case the modulation at Ω gives the Pockels coefficient and that at 2Ω the Kerr one. Similarly as explained in the section concerning refractive index measurement, it is possible to determine the electro-absorption (modulation of the imaginary part of refractive index κ) by a careful fit of the shape of the modulation of the reflection deeps.

A characteristic signature of ferro-electric films is the sign change of modulation between the modes of very low order and those close to the cut off (Dentan, 1989). The reason is that the effective index of small order modes are much more dependent on the film refractive index that on its thickness. Consequently they increase with thickness decrease, according to Eq. (71). On the opposite hand, modes of higher orders close to the cut off are much more sensitive to the thickness variation and their effective index decrease with the thickness. This results in an inversion of the modulation shape, as shown in Figure 9.

6.2.2.4   Ellipsometric method

This method takes opportunity of the electro-optic tensor anisotropy in poled films which originates from the fact that the nonlinear chromophores can be consider as

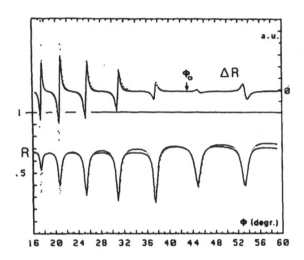

Figure 9 Electric field induced Modulation of ATR (cf. Figure 6) for a piezo-electric film. The effective index derivatives of the low order modes (incidence angle larger than Φ_0) are predominantly governed by the film refractive index variation (electro-optic phenomenon) while for the high order modes (incidence angle smaller than Φ_0) the influence of the film thickness variation (piezo-electricity) is dominant (after Dentan *et al.*, 1989).

linear molecules and because they are poled in the direction normal to the plane (Schildkraut, 1990; Teng, 1990). It results that under the application of an electric field in the direction normal to the plane (labeled 3) the extraordinary index is more modulated than the ordinary one (3 times in the gas phase model). The consequence is that the phase of a monochromatic beam polarized in the incidence plane (*p*-polarization) is more modulated than for a beam polarized in the perpendicular direction (*s*-polarization). This phase modulation mismatch is converted into an intensity modulation thanks to the experimental set up described in Figure 10. The film is sandwitched between a transparent electrode and a reflecting one. The incident beam is polarized at 45° from the incidence plane. In addition a Bravais–Soleil–Babinet (BSB) compensator enables to vary the phase mismatched of the *s* and *p* polarization from a value Ψ_B.

Simplified theory. In this approximation it is assumed that the phase mismatch originates only from the electro-optic film, and the multiple reflections and losses are neglected (Schildkraut, 1990). The reflected intensity is given by

$$I_r = I_i \sin^2(\Psi_{ps} + \Psi_B) \tag{88}$$

Ψ_{ps} is the phase mismatch inside the film between *s* and *p* waves given by

$$\Psi_{ps} = \frac{4\pi h}{\lambda}\left\{\frac{n_o}{n_e}(n_e^2 - n_a^2\sin^2\theta_a)^{1/2} - (n_o^2 - n_a^2\sin^2\theta_a)^{1/2}\right\} \tag{89}$$

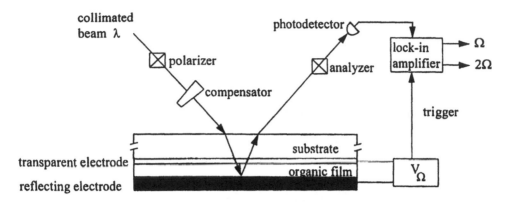

Figure 10   Experimental setup for measuring the electro-optic coefficients by the ellipsometric method.

When the setting of the BSB compensator is varied, the reflection modulation varies according to

$$\frac{\partial I_r}{\partial E} = \sin\{2(\Psi_{ps} + \Psi_B)\}\frac{\partial \Psi_{ps}}{\partial E}.$$

(90)

It reaches its maxima for $\Psi_{ps} + \Psi_B = \pm\pi/2$ (points A and B in Figure 11).

Calculating $\partial \Psi_{ps}/\partial E$ from Eq. (89) one can deduce in the case of the small anisotropy of the film ($n_o \approx n_e = n$) the connection between the eletro-optic coefficients and the reflection modulationI_m^A at point A:

$$\frac{I_m^A}{I_i} = \frac{F}{2}r_{33}V_m,$$

(91)

$$F = \frac{4\pi}{3\lambda}\frac{n^2 n_a^2 \sin^2 \theta_a}{(n^2 - n_a^2 \sin^2 \theta_a)^{1/2}}.$$

(92)

We have assumed a ratio 3 between r_{33} and r_{13} (gas phase model) and n is the mean refractive index of the film. In the case of slightly absorbing films, this results can be extended to electro-absorption (Lévy, 1993) when the Pockels coefficient is complex ($r = r^r + r^i$):

$$r_{33}^r = \frac{I_m^A - I_m^B}{F I_i V_m},$$

(93)

$$r_{33}^i = \frac{I_m^A + I_m^B}{F I_i V_m}.$$

(94)

Rigorous theory. The reflection coefficients of the whole stratified structure is complex and can be written

$$r_p = \rho_p e^{i\Psi_p},$$ (95)

$$r_s = \rho_s e^{i\Psi_s}.$$ (96)

The intensity detected after crossing the analyzer is equal to

$$I_d = \frac{I_i}{4}\left| r_p e^{i\Psi_s} - r_s \right|^2 = \frac{I_i}{4}\{\rho_p^2 + \rho_s^2 - 2\rho_p\rho_s \cos(\Psi_{ps} + \Psi_B)\}$$ (97)

which reaches its maximum $I_{max} = (\rho_p + \rho_s)^2 \frac{1}{4}$ at point O and its intermediate values at points A and B (see Figure 11). Denoting $I_m^{A,B,O}$ the derivatives $\partial I_d^{A,B,O}/\partial E$ at points A, O and B it is strightforward to deduce:

$$\frac{\partial \Psi_{ps}}{\partial E} = \frac{\partial \varphi_p}{\partial E} - \frac{\partial \varphi_s}{\partial E} = \frac{I_m^A - I_m^B}{\rho_p \rho_s I_i},$$ (98)

$$\frac{\partial \rho_p^2}{\partial E} - \frac{\partial \rho_s^2}{\partial E} = 2\frac{I_m^A - I_m^B}{I_i}.$$ (99)

In addition, contrarily to the prediction of the simplified model, there is a modulation signal at the O point (maximum of reflection) given by

$$I_m^O = \frac{I_i}{4}\frac{\partial(\rho_p + \rho_s)^2}{\partial E}.$$ (100)

Equations (98) and (99) play asymmetrical role, the first dealing with the phase mismatch (real part of the electro-optic effect) and the second with the amplitude (imaginary part of the electro-optic effect). In fact owing to the fact that multiple reflection and losses occur in the different layers, the real and imaginary part of the refractive index modulation (including its anisotropy) can be precisely obtained only from a fit of the reflection modulation under variable incidence angle (Lévy, 1993; Chollet, 1994) as shown in Figure 12. The quadratic electro-optic effect has also been studied (Chollet et al., 1995).

This ellipsometric method has the advantage of being an experimental method quite easy to set up and the deduction of the electro-optic coefficients from Eqs. (91–92) is straightforward out of absorption, in the gas phase model assumption ($r_{13} = r_{33}/3$). However working inside the absorption range needs more thorough interpretation.

6.2.2.5 *Grating diffraction*

This technique (Shi et al., 1994) consists in etching the electro-optical in order to make a grating. The grating is sandwiched between two transparent electrodes. A laser beam which illuminates the grating at normal incidence is deflected into the various orders. The deflection angles are governed by the period of the grating, and the intensity of each order depends on the grove depth of the grating and the

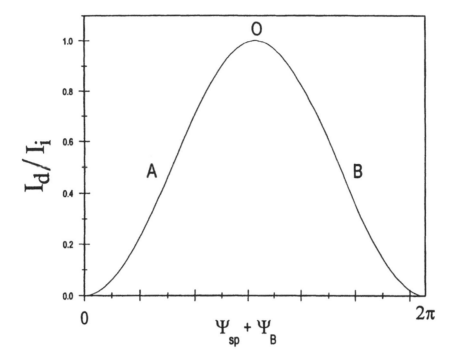

Figure 11 Ellipsometric method: reflected intensity variation versus total phase mismatch $\Psi_{ps} + \Psi_B$ between TM(p) and TE(s) polarizations, including the contribution of the Soleil–Bravais–Babinet compensator. In the simplified model, points A and B corresponds to the maxima of the modulation signal.

refractive index of the film. Consequently an electromodulation of the refractive index generates a modulation of the various order intensities. It is possible to deduce the electro-optic and electro-absorption coefficients from the intensity modulation of the different diffracted orders.

7 APPLICATIONS

7.1 Frequency conversion methods

As already mentioned one of the important goal targeted with noncentrosymmetric thin film is frequency conversion. Several more or less succesful approaches have been used up to now. The most evident is the phase matching technique.

7.1.1 Phase matching. As already mentioned in the second harmonic generation process two harmonic waves propagate in the nonlinear medium: free wave with the velocity of harmonic wave and bound wave with the velocity of fundamental wave. Due to the refractive index dispersion both velocities are different and as consequence

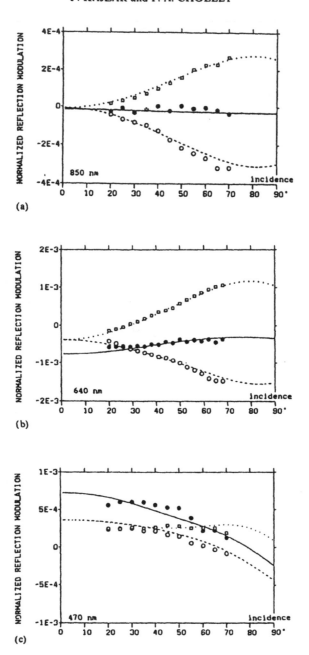

Figure 12 Ellipsometric method: comparison between the experimental results for different setting of the compensator (squares: point A; open circles: point B; full circles: point O; cf. Figure 11) and the theoretical values (dots, broken and solid curves) calculated using the rigorous model. The experiments were made far from absorption (a: $\lambda = 850$ nm) at maximum absorption (c: $\lambda = 470$ nm) and for an intermediate wavelength (b: $\lambda = 640$ nm). After Chollet *et al.* (1994).

there is a phase mismatch between two waves. Consequently interference of both waves leads to Maker fringes with characteristic maximas and minimas. In a dispersionless medium both waves will interfer constructively along the whole propagation length leading to a maximum conversion. In anisotropic materials refractive index depends on the propagation direction and it is sometimes possible to realize the phase matching by choosing the propagation direction in such a way that refractive indices of harmonic and fundamental waves are equal. Practically two configurations are possible, selected by a corresponding choice of fundamental and harmonic field polarizations: $o + o \rightarrow e$ (or $e + e \rightarrow o$ for negative crystals, phase matching I) or $o + e \rightarrow e$ (or o) (phase matching II), where "o" refers to ordinary and "e" to extraordinary rays, respectively. Although this is feasible with bulk crystals, it is impossible with poled polymer films (Burland, 1994). In order to overcome the difficulties with realization of phase matching in thin films with point symmetry ∞ mm polymer films several other structures have been invented which are described below.

7.1.2 Quasi phase matching. As already mentioned the free and bound wave interfere constructively on the distance equal to the coherence length l_c. In order to get a constructive interference on the propagation length between l_c and $2l_c$ (or in general between $(2_j - 1)l_c$ and $4_j l_c$, where $j = 1,2, \ldots$) it is sufficient to inverse the direction of the diagonal $\chi^{(2)}$ tensor component. This has been made succefully with ferroelectric crystals like $LiNbO_3$ or KTP by a periodic ion etching and a subsequent domain inversion taking advantage of different ferroelectric transition temperatures in etched and nonetched areas.

In the case of organic materials this kind of structure is more difficult to realize. However two different structures have been proposed:

(i) use of finger electrodes with alternating direction of electric field, as proposed by Khanarian *et al.* (1990, 1996). The length of electrode is almost equal to the coherence length. It applies obviously to poled polymers. The conversion efficiency which is defined as

$$\eta = P(2\omega)/[P(\omega)]^2 \tag{101}$$

and obtained with such a type of structure is limited due to different encountered problems like: chromophore orientation and temperature stability, orientation of chromophores in the inter-electrode areas, electric field profile, etc. . . (Bosshard *et al.*, 1995). Khanarian *et al.*, 1996 have proposed recently another possibility of quasi phase matching realization with poled polymers by stacking thin films with oppositely directed dipolar moments and thickness equal to the coherence length. This can be done by polarizing a large area thin film, detaching it from the substrate and cutting into slabs which can be stacked with desired order.

7.1.3 Modal phase matching. In an optical waveguide, depending on its thickness and refractive index as well that of substrate a certain number of TE and TM modes can be excited. As refractive index dispersion for different modes is different it is possible to find a fundamental mode (TE or TM) and a harmonic (TE or TM) with the same refractive indices. However different modes have different electric field distribution. Consequently the conversion efficiency depends on the overlap integral between modes under question and for a light beam propagating in x-direction it is given by

$$S = \left| \iint_{-\infty}^{\infty} E_m^2(\omega,y,z) E_n(2\omega;y,z) dy dz \right|^2 \tag{102}$$

where m and n refer to the fundamental and harmonic modes, respectively.

7.1.4 Cerenkov conversion. Usually the effective refractive index of harmonic wave is larger than that of fundamental. However by an adequate choice of substrate refractive index and thin film thickness it is possible to guide only the fundamental mode and that the harmonic one is emitted through the substrate (Figure 13). This is the so called Cerenkov type conversion. The condition for effective refractive indices for thin film (f) and substrate (s) to be satisfied are the following ones (Chikuma *et al.*, 1990):

$$n_s^{2\omega} > n_{\text{eff}}^{\omega} > n_s^{\omega} \tag{103}$$

where ω and 2ω refer to fundamental and harmonic beam and n_{eff} is the effective index of refraction of waveguide at frequency ω. The harmonic conversion efficiency with such a type of structure is limited due to the limited distance at which both free and bound waves interfere (cf. Eqs. 28 and 29 in Chikuma *et al.*, 1990).

7.1.5 Counter propagating beam SHG. Another possibility of frequency conversion has been proposed by Stegeman and coworkers (Normandin, 1979; 1980; 1982; 1991; Stegeman, 1987; Otomo, 1995). By sending two short laser pulses into a channel waveguide and in opposite direction (z), which can be described as (cf. Otomo, 1995):

$$E_\pm^\omega = \tfrac{1}{2}[\varepsilon_\pm(x) e^{i(\omega t \pm \beta^\omega z)} + c \cdot c], \tag{104}$$

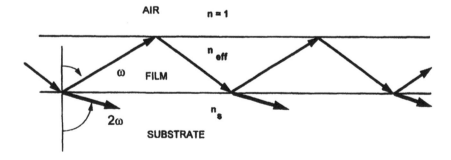

Figure 13 Schematic representation of Cerenkov type experiment configuration for second harmonic generation in an organic waveguide. Due to an adequate choice of the refractive index relative magnitude of the film and substrate, the fundamental wave is guided, whereas the second harmonic wave radiates into substrate.

where $\varepsilon_{\pm}$ is the transverse guided field (cf. Figure 14), β^{ω} is the propagation constant of the fundamental guided wave, The second harmonic is emitted in the direction perpendicular to the substrate and it is given by the following equation:

$$I^{2\omega} = \varepsilon_0 \omega k_0^{2\omega} n_c^{2\omega} |C|^4 |S(\infty)|^2 I_+^\omega I_-^\omega, \tag{105}$$

where I are fundamental beam intensities, C is the normalization constant of the guided mode profile and $S(\infty)$ is the overlap integral

$$S(\infty) = \int_{-\infty}^{\infty} \frac{d_{22}(x')f_y(x')^2 e^{ik_0^{2\omega} n_f^{2\omega} x'}}{n_f^{2\omega}} dx' \tag{106}$$

where $f_y(x)$ is the guided mode profile. To optimize the conversion efficiency it is important to control well the thin film thickness (cf. Otomo, 1995).

7.2 Electro-optic Modulation

Another important application targeted with noncentrosymmetric thin films is electro-optic modulation for signal transmission. The simplest structure is a phase modulator, shown schematically in Figure 15. In order to get the amplitude modulation two types of structure have been proposed:

(i) Mach–Zehnder interferometer, shown schematically in Figure 16. By applying an electric field to one branch of interferometer one creates a phase mismatch between waves propagating in both branches and consequently a modulation of light intensity at the output of interferometer. Usually the electric field is applied to both arms of interferometer what permit its better balancing. A modulation

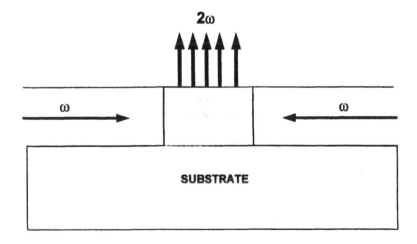

Figure 14 Schematic representation of second harmonic generation in the counter propagating wave configuration.

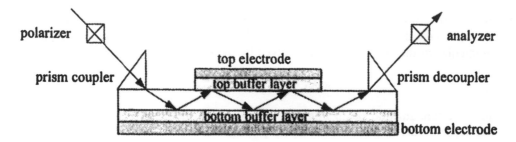

Figure 15 Schematic representation of a polymer phase modulator. The polarizer an analyzer, oriented in a parallel direction, allow to select the *s* or *p* polarization. The analyzer is not necessary (if no polarization coupling is present).

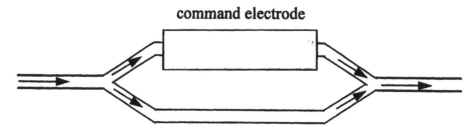

Figure 16 Schematic representation of an integrated Mach–Zehnder electro-optic intensity modulator.

rate as high as 94 GHz has been reported recently with poled polymer based Mach–Zehnder interferometer (Dalton *et al.*, 1996).

(ii) Digital modulator (cf. Figure 17.) Using a Y type structure with an electroded arm it is possible to control the light intensity distribution in output arms by the control of applied voltage. If the structure is made in such a way that at zero voltage light goes through channel 2, by modification of refractive index, through the applied external field one can switch the light continuosly from channel 2 to channel 1. Consequently the light intensity outgoing from channel 1 will depend on the applied voltage, as shown in Figure 18. Such a structure can be used for light intensity modulation or for switching.

8 CONCLUSIONS

In this paper we have reviewed noncentrosymmetric organic thin film preparation techniques, their linear and nonlinear optical properties characterization methods as well as some applications.

Between the discussed here oriented film preparation methods poled polymers exhibit the most important progress, concerning the obtained values of second order NLO susceptibilities, thermal and orientational stability as well as light propagtion properties. One can reasonably expect their forthcoming, commercial application in

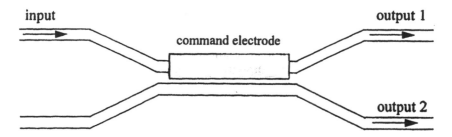

Figure 17 Schematic representation of a directional (digital) electro-optic modulator. It can be used as an intensity modulator (cf. Figure 18) or for beam steering.

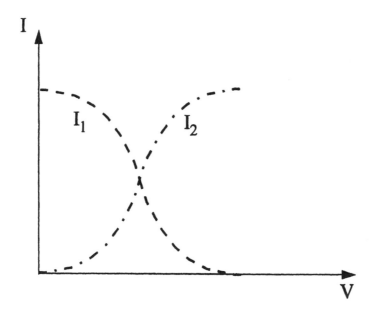

Figure 18 Idealized voltage dependence of the light intensity in the two ports of the directional coupler shown in Figure 17.

high rate electro-optic modulation. Better poling efficiency is also expected with photo-assisted and all optical poling. In particular, all optical poling technique has several important advantages, with respect to the classical static field poling method. These include:

(i) the absence of electrodes and consequently no charge injection, which usually leads to dielectric breakdown

(ii) a higher damage threshold observed with optical fields E, allowing consequently higher poling efficiencies, with no "point effect", observed sometimes with electrodes

(iii) automatic phase matching, assured by the seeding procedure and useful for the frequency conversion

(iv) possibility of autoregeneration, as frequency conversion is done simultaneously with seeding. It allows an automatic $\chi^{(2)}$ grating adjustment for harmonic conversion to e.g. refractive index variation due to e.g. heating

 (v) poling can be done at any temperature, especially at room temperature ("cold poling") or at the operation temperature

(vi) possibility of poling other than dipolar molecules as it was shown with ocupolar molecules. Although the poling mechanism is still not clear, it is the only technique which allows the orientation of these interesting molecules.

We have discussed two applications of oriented thin films:

– frequency conversion

– electro-optic modulation.

The most important progress has been noticed with electro-optic modulation. As already mentioned, researchers from University of South California have demonstrated electro-optic modulation at 94 GHz rate. Time resolved experiments (Ferm *et al.*, 1991) show that the rise on time with poled polymers is of 450 femtoseconds allowing a modulation rate of 2 THz. As a matter of fact the theoretical limit lies at still higher frequencies. The electronic origin of the NLO response and the good relationship between SHG and EO coefficients allows to place the modulation rate at optical frequencies (in SHG process the harmonic field follows the fundamental one), the practical limitations coming from the difficulties related to the application of external high frequency field to the modulator, what requires a special shape of electrodes in order to avoid they work as antennas.

Less progress have been made with frequency conversion, as compared with inorganic quasi phase matched structures, based on KTP or $LiNbO_3$. Difficulties come from different sources like:

– practical realization of phase matched structures.

– photochemical stability.

It is well known that the nonlinear optical response of organic materials depends strongly on the conjugation length, and consequently on optical transition energy. With second harmonic light close to the absorption band one can expect photodegradation of molecules and consequently a loss of nonlinear response as well as phase matching due to the associated refractive index variation. Photo induced damage can come also from higher order nonlinear optical processes. A very little study has been done up to know in order to quantify this problem.

We did not discuss another interesting application of this kind of materials which is linear rectification. This effect is useful to create very short electrical pulses, with duration equal to the pulse duration. By a use of femtosecond laser sources one can expect generation of THz pulses.

ACKNOWLEDGEMENTS

Thanks are due to Maryanne Large for careful reading and correcting the manuscript.

REFERENCES

Abelès, F., *J. Phys. Radium* **11**, 310 and 403 (1950).

Aktsipetrov, O.A., Akhmediev, N. N., Mishina, N. N. and. Novak, V. R., *Soviet Phys. JETP Lett.* **37**, 207 (1983).

Azzam, R.M.A. and Bashara, N.M., Ellipsometry and polarized light, North-Holland publishers, Amsterdam (The Netherlands) (1979).

Boccara, A.C., Fournier, D., Jackson, W.B. and Amer, N.M., *Opt. Lett.* **5**, 377 (1980).

Bosshard, C., Flörsheimer, M., Küpfer, M. and Günter, P., *Opt. Commun.* **85**, 247 (1991).

Bosshard, C., "Oriented Molecular Systems", in "Organic Thin Films for Waveguiding Nonlinear Optics", F. Kajzar and J. Swalen Eds., Gordon and Breach Sci. Publ., Amsterdam (1996), pp. 163–191.

Boyd, G.T., Francis, C.V., Trend, J.E. and Ender, D.A., *J. Opt. Soc. Am. B* **8**, 887 (1991).

Burland, D.M., Walsh, C.A., Kajzar, F. and Sentein, C., *J. Opt. Soc. Am. B* **8**, 2269 (1991) .

Burland, D.M., Miller, R.D. and Walsh, C.A., Chemical Reviews **94**, 31(1994).

Charra, F., Kajzar, F., Nunzi, J.M., Raimond, P. and Idiart, E., *Opt. Lett.* **18**, 941 (1993).

Charra, F., Fiorini, C., Kajzar, F., Nunzi, J. M., Raimond, P. and Chapulczak, W., "All Optical Poling of Polymers" in "Photoactive Organic Materials", F. Kajzar, V. M. Agranovich and C. Y. Lee Eds, NATO ASI SERY 3, vol. 9, Kluwer Academic Publ., Dordrecht (1996), pp. 513–526.

Chikuma, K. and Umegaki, S., *J. Opt. Soc. Am. B* **7**, 768 (1990).

Chollet, P.-A., Kajzar, F. and Le Moigne, J., *Mol. Engineering* **1**, 35 (1991).

Chollet, P.-A., Gadret, G., Kajzar, F., Raimond, P. and Zagorska, M., *Proc. SPIE* **1775**, 121 (1992).

Chollet, P.-A., Gadret, G., Kajzar, F. and Raimond, P., *Thin Solid Films* **242**, 132 (1994).

Chollet, P.-A., Gadret, G. Kajzar, F. and Raimond P., *Proc. of Iketani Conf.*, Hawaii 1994, *Nonl. Opt.* **14**, 47 (1995).

Chollet, P.-A. and Lévy, Y., "Characterization Techniques of Nonlinear Optical Thin films" in "Organic Thin films for Waveguiding Nonlinear Optics", F. Kajzar and J.D. Swalen Eds., Gordon and Breach Publ., Amsterdam (1996), pp. 457–503.

Choy, M. and Byer, R.L., *Phys. Rev. B* **14**, 1693 (1976).

Clays, C. and Persoons, A., *Phys. Rev. Lett.* **66**, 2980 (1991).

Clays, C. and Persoons, A., *Rev. Sc. Instrum.* **63**, 3285 (1992).

Comizzoli, B., *J. Electrochem. Soc.: Solid Science and Technology* **134**, 424 (1987).

Coutaz, J.L., Blau, G., Roux, J.-F., Reinisch, R., Kajzar, F., Raimond, P., Robin, P., Chastaing, E. and Le Barny, P., *Nonl. Opt.* **10**, 347 (1995).

Cyvin, S.J., Rauch, J.E. and Decius, J.C., *J. Chem. Phys.* **63**, 3285 (1965).

Dantas-Demorais, T., Noel C. and Kajzar, F., *Proc. of ICONO'2, Nonl. Opt.* **15**, 315 (1996).

Dentan, V., Lévy, Y., Dumont, M., Robin, P. and Chastaing, E., *Opt. Comm.* **69**, 379 (1989).

Dumont, M., Lévy, Y. and Morichère, D. in "Organics Molecules for Nonlinear Optics and Photonics", J. Messier, F. Kajzar and P. Prasad Eds., NATO ASI Series E, Vol. 194 p. 461, Kluwer Academic Publishers, Dordrecht (1991).

Dumont, M., Froc, G. and Hosotte, S. *Nonl. Opt.* **9**, 327 (1995).

44 F. KAJZAR and P.-A. CHOLLET

Ederling, C.A., Kowell, S.T. and Knoessen A., *Appl. Optics* **28**, 4442 (1989).
Ferm, P.M., Knapp C.W., Wung C., Yardley, J.T., Hu, B.-B., Zhang, X.-Ch. and Auston, D.H., *Appl. Phys. Lett.* **59**, 2651 (1991).
Franke, H., Knabke, G. and Reuter, R., *Proc. SPIE* **682**, 191 (1986).
Gadret, G., Kajzar, F. and Raimond, P., *Proc. SPIE* **1560**, 226 (1991).
Gonin, D., Noël C. and Kajzar, F. *Nonl. Opt.* **8**, 37 (1994a).
Gonin, D., PhD Thesis, Université Pierre et Marie Curie, Paris, (1994b).
Gonin, D., Noël, C. and Kajzar, F., "Liquid crystalline polymers", in "Organic Thin Films for Waveguiding Nonlinear Optics". F. Kajzar and J.D. Swalen Eds., Gordon and Breach Publ., Amsterdam (1996), pp. 221–288.
Heavens, O.S., Thin Film Physics, Methuen and Co Publ., London (1970) p. 65.
Hoshi, H., Nakamura, N., Maruyama, Y., Nakagawa, T., Suzuki, S., Shiromaru, H. and Achiba, Y., *Jap. J. Appl. Phys.* **30**, L1397 (1991).
Hunsperger, R.G., Integrated Optics; Theory and Technology, Springer-Verlag, Berlin (1982).
Kajzar, F., Ledoux, I. and Zyss, J., *Phys. Rev. A* **36**, 2210 (1987).
Kajzar, F. and Ledoux, I., *Thin Solid Films* **179**, 359 (1989).
Kajzar, F., Chollet, P.A., Ledoux, I, Lemoigne, J., Lorin, A. and Gadret, G., "Organic thin films for quadratic optics", in: "Organic Molecules for Nonlinear Optics and Photonics", NATO ASI Series, Series, E, Vol. 194, pp. 403–432, Kluwer Academic Publ., Dordrecht, (1991).
Kajzar, F., Taliani, C., Zarnboni, R., Rossini, S. and Danieli, R., *Synth. Metals* **54**, 21 (1993).
Kajzar, F., Horn, K., Nahata, A. and Yardley J.T., *Nonl. Opt.* **8**, 205 (1994).
Kajzar, F., Charra, F., Nunzi, J.M., Raimond, P., Idiart, E. and Zagorska, M., *Proceedings of International Conference on Polymers and Advanced Materials*, Jakarta, January 1993, P.N. Prasad Ed., Plenum (1995a).
Kajzar, F. and Zyss, J., *Nonl. Opt.* **9**, 3 (1995b).
Khanarian, G. and Norwood, R.A., *Proc. SPIE* **1337**, 44 (1990).
Khanarian, G., Mortazavi, M. and Norwood, R.A., in "Organic Thin Films for Waveguiding Nonlinear Optics: Science and Technology", F. Kajzar and J. Swalen Eds., Gordon and Breach Publ., Amsterdam (1996) pp. 689–757.
Koopmans, B., Anema, A., Jonkman, H.T., Sawatzky, G.A. and Van der Woude, F.X, *Phys. Rev. B* **48**, 2759 (1993).
Kobayashi, T., Minoshima, K., Nomura, S., Fukaya, S. and Ueki, A., *Proc. SPIE* **1147**, 182 (1989a).
Kobayashi, T., Uchiki, H. and Minoshima, K., in: "Nonlinear Optical Properties of Organics and Semiconductors" T. Kobayashi, Ed., *Springer Proceedings in Physics* **36**, 140 (1989b).
Küpfer, M., Flörsheimer, M., Bosshard, C., Looser, H. and Günter, P., *Proc. SPIE* **1775**, 340 (1994).
Kuzyk, M.G., Moore, R.C. and King, L.A., *J. Opt. Soc. Am. B* **7**, 64 (1990a).
Kuzyk, M.G., Sohn, J.E. and Dirk, C.W., *J. Opt. Soc. Am. B* **5**, 842 (1990b).
Le Moigne, J., Kajzar, F. and Thierry, A., *Macromol.* **24**, 2622 (1991).
Le Moigne, J., Oswald, L., Kajzar, F. and Thierry, A., *Nonl. Opt.* **9**, 187 (1995).
Levine, B.F. and Bethea, C.G., *J. Chem. Phys.* **63**, 2666 (1975).
Levy, Y., Dumont, M., Chastaing, E., Robin, P., Chollet, P.-A., Gadret, G. and Kajzar, F., *Nonl. Opt.* **4**, 1 (1993).
Maker, P.D., *Phys. Rev. A* **1**, 923 (1970).
Morichère, D., Dentan, V., Kajzar, F., Lévy, Y. and Dumont, M., *Opt. Comm.* **69**, 774 (1989).

Mortazavi, M.A., Knoesen, A., Kowel, S.T., Higgins, B.G. and Dienes, A., *J. Opt. Soc. Am.* B **6**, 733 (1989).

Normandin, R. and Stegeman, G.I., *Opt. Lett.* **4**, 58 (1979).

Normandin, R. and Stegeman, G.I., *Appl. Phys. Lett.* **36**, 253 (1980).

Normandin, R. and Stegeman, G.I., *Appl. Phys. Lett.* **40**, 759 (1982).

Normandin, R., Letorneau, S., Chatenoud, F. and Williams, R.L., *J. Quant. Electr.* **QE-27**, 1520 (1991).

Norwood, R.A., Findakly, T., Godberg, H.A., Khanarian, G., Stamatoff, J.B. and Yonn, H.N., "Optical Polymers and Multifunctional Materials" in "Polymers for Lightwave and Integrated Optics" L.A. Hornak Ed., Marcel Dekker Publisher, New York (1992).

Nunzi, J.M., Charra, F., Fiorini, C. and Zyss, J., *Chem. Phys. Lett.* **219**, 349 (1994).

Nunzi, J.M., Fiorini, C., Charra, F., Kajzar, F. and Raimond, P., "All-Optical Poling of Polymers for Phase-Matched Frequency Doubling, in Polymers for Second-Order Nonlinear Optics", G.A. Lindsay and K.D. Singer Eds., ACS Symposium Series, Vol. 601, p. 240, ACS, Washington (1995).

Onsager, L., *J. Am. Chem. Soc.* **58**, 1486 (1936).

Orr, B.J. and Ward, J.F., *Mol. Phys.* **20**, 513 (1971).

Osterberg, H. and Smith, L.W., *J. Opt. Soc. Am.* **54**, 1078 (1964).

Otomo, A., Bosshard, C., Mittler-Neher, S., Stegeman, G.I., Küpfer, M., Flörsheimer, M., Günter, P., Horsthuis, W.H.G. and Möhlmann, G.R., *Nonl. Opt.* **10**, 331 (1995).

Oudar, J.L. and Chemla, D.S., *Opt. Commun.* **13**, 10 (1975).

Oudar, J.L., *J. Chem. Phys.* **67**, 446 (1977).

Page, R.H., Jurich, M.C., Reck, B., Sen, A., Twieg, R.J., Swalen, J.D., Bjorklund, G.C. and Wilson, C.G., *J. Opt. Soc. Am. B* **7**, 1239 (1990).

Penner, T.L., Motschmann, H.R., Armstrong, N.J., Ezenyilimba, M.C. and Williams, D.J., *Nature* **367**, 49 (1994).

Röhl, P., Andress, B. and Nordmann, *J. Appl. Phys. Lett.* **59**, 2793 (1991).

Sckkat, Z. and Dumont, M., *Nonl. Opt.* **2**, 359 (1992).

Shen, Y.R., *Nature* **337**, 519 (1989).

Schildkraut, J.S., *Appl. Opt.* **29**, 2839 (1990).

Shi, Y., Bechtel, J.H., Kalluri, S., Steier, W., Xu, C., Wu, B. and Dalton, L. R. *Proc., SPIE* **2285**, 131 (1994).

Singer, K.D. and Garito, A.F., *J. Chem. Phys.* **75**, 3572 (1981).

Skumanich, A., Jurich, M. and Swalen, J.D., *Appl. Phys. Lett.* **62**, 446 (1993).

Stegeman, G.I., Burke, J.J. and Seaton, C.T., "Nonlinear Integrated Optics", in Optical Engineering Integrated Optical Circuits and Components, Hutcheson, L.D. Ed., Marcel Dekker, New York (1987).

Sugihara, T., Haga, H. Yamamoto, S., *Appl. Phys. Lett.* **68**, 144 (1996).

Swalen, J. and Kajzar, F., in "Organic Thin Films for Waveguiding Nonlinear Optics: Science and Technology", F. Kajzar, and J. Swalen Eds., Gordon and Breach Sci. Publ., Amsterdam (1996) pp. 1–44.

Teng, C.C. and H.T. Man, *Appl. Phys. Lett.* **56**, 1734 (1990).

Tien, P.K., Ulrich, R. and Martin, R.J., *Appl. Phys. Lett.* **14**, 291 (1969).

Terhune, R.W., Maker, P.D. and Savage, C.M., *Phys. Rev. Lett.* **14**, 681 (1965).

Uchiki, H. and Kobayashi, T., *J. Appl. Phys.* **64**, 2625 (1988).

Verbiest, T., Hendrickx, E. and Persoons, A., *Proc. SPIE* **1775**, 206 (1993).

Wang, X.K., Zhang, T.G., Lin, W.P., Shen Zhong Liu, Wong, G.K., Kappes, M.M., Chang, R.P.H. and Ketterson, J.B., *Appl. Phys. Lett.* **60**, 810 (1992).

Weber, H.P., Dunn, F.A. and Leibolt, W.N., *Appl. Opt.* **12**, 755 (1973).

Yamada, T., Hoshi, H., Ishikawa, K., Takezoe, H. and Fukuda, A., *Proc. of ICONO'2, Nonl. Opt.* **15**, 131 (1996).

Zernike, F., Douglas, J.W. and Olson, D.R., *J. Opt. Soc. Am.* **61**, 678 (1971).

Zyss, J. and Oudar, J.L., *Phys. Rev. A* **26**, 2028 (1982).

Zyss, J., *Nonl. Optics* **1**, 3 (1991).

2. THERMALLY STABLE SECOND-ORDER NONLINEAR OPTICAL POLYMERS

NAKJOONG KIM[a], DONG HOON CHOI[b], and SOO YOUNG PARK[c]

[a] Division of Polymer Research, Korea Institute of Science and Technology, P.O. Box 131, Cheongryang, Seoul 130-650, Korea; [b] Department of Textile Engineering, Kyung Hee University, Yongin-Kun, Kyungki-Do, 449-701, Korea; [c] Department of Fiber and Polymer Science, Seoul National University, Seoul 151-742, Korea

ABSTRACT

Second-order nonlinear optical copolymers containing methylmethacrylate (MMA), p-hydroxyphenyl maleimide (HPMI) or glycidylmethacrylate (GMA) as a comonomer were newly synthesized. Second-order nonlinear optical properties of the poled or poled/cured films were investigated in terms of the second harmonic generation (SHG) and linear electro-optic (EO) coefficient measurement. Second-order NLO coefficient, d_{33} and EO coefficients, r_{33} were also measured at 1064 and 632 nm wavelength, which were significantly high. To prevent the molecular relaxation of poled polymer, thermal crosslink reaction was induced either using diisocyanate as a crosslinker or by virtue of self-crosslinking reaction between the side chains themselves. Temporal stabilities of second-order NLO coefficients of crosslinked polymer systems were proved much better than those of other side chain polymers.

INTRODUCTION

Organic nonlinear optical (NLO) materials provide strong potential advantages for second harmonic generation and electro-optic applications [1–4]. Particularly, side-chain polymers have drawn remarkable interest in recent years as promising candidates for application in electro-optic and photonic devices [1–6]. Second-order NLO properties of poled polymers have been extensively studied in the past. The important issue for practical application with poled polymers is postulated that temporal stability of dipolar alignment should be improved. For this purpose, two methods look quite promising: one is to synthesize the thermoplastic polymer whose glass transition temperature is very high; and the other is to prepare the crosslinked polymer system [7–9]. Although both methods have been proven to be effective preventing the dipolar relaxation, it should be pointed out that achieving a higher level of molecular alignment, namely a high second-order NLO coefficient, is very difficult or limited with these methods.

In the past studies, the crosslinked system was also very successful to achieve the better temporal stability compared to the pure copolymer. Number of studies were carried out to improve the temporal stabilities using crosslink reaction so far. Most of the polymer backbone were considered to be polyacrylate or polystyrene in the category of the side chain polymer. Besides these soft backbone structures, more rigid structures would be necessary to be taken into account for our purpose. Adding the crosslinking agent into the mother solution, the optical quality film will be corrupted frequently, particularly after poling and curing process. In addition to the heterogeneous crosslinking systems, another novel system should be required to develop subjugating the above shortcomings. Crosslinkable unit can be introduced into the inherent polymer structures to induce spontaneous crosslink either between the side chains or between the main chain and the side chain.

In an attempt to improve both the second-order NLO effect and its temporal stability, we designed and synthesized the copolymers containing the crosslinkable unit in the side chain of the comonomer. These polymers are easy to synthesize and quite soluble in common organic solvents. We adopted the method of thermal crosslink reaction between the side chains themselves. Two different methods for thermal crosslink were employed using heterogeneous system containing the copolymer with diisocyanate and homogeneous system of copolymer itself. After we have synthesized novel copolymers containing either the piperazyl nitrostilbene (CP-HPMI) or methylaminonitrostilbene (CP-GMA) chromophore in the side chain, the second-order NLO properties of new crosslinked polymers were investigated by virtue of the second harmonic generation and linear electro-optic coefficient measurement.

EXPERIMENTAL

Synthesis of Monomer and Polymer

4-[N-(2-hydroxyethyl)piperazyl]benzaldehyde (I) and 4-[N-(2-hydroxyethyl)piperazyl]-4'-nitrostilbene (II) were prepared by reported methods [10]. The syntheses of α-methylstyrene analogue and its copolymer with MMA are illustrated in Scheme 1.

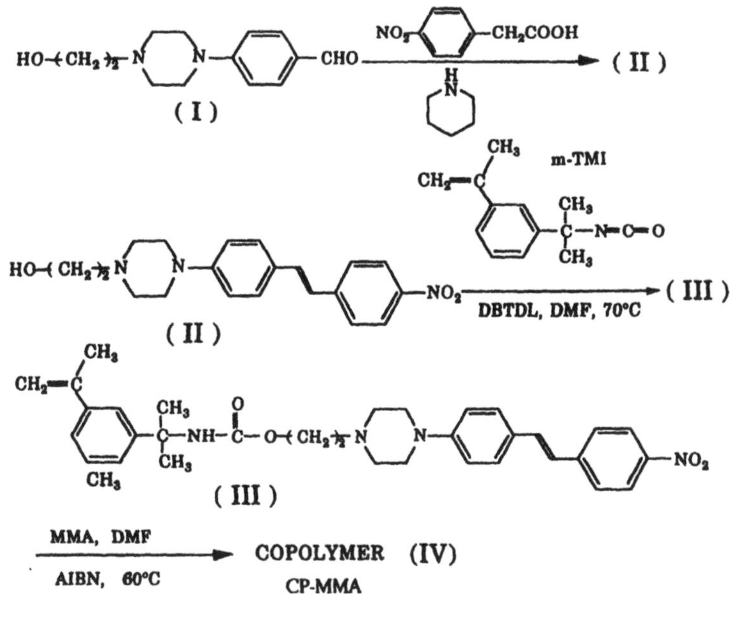

Scheme 1   Synthetic procedure of CP-MMA.

p-hydroxyphenyl maleimide: HPMI (V) was obtained from the reaction with maleic anhydride and *p*-aminophenol by the literature method [11]. The monomer containing NLO chromophore in its side chain, 4'-[(methacryloxyethyl) methyl-amino]-4-nitrostilbene was prepared according to the known procedure [12].

[1-[4-(N-ethylenepiperazyl)-4'-nitrostilbene]-N-[(1,1-dimethyl-*m*-isopropenyl)benzyl]]carbamate (III)

4-[*N*-(2-hydroxyethyl)piperazyl]-4'-nitrostilbene (2.9 g, 8.21 mmole) was dissolved in dry dimethylformamide (DMF, 200 ml) at 25°C under nitrogen atmosphere. Then, a trace of dibutyltin dilaurate (0.15 mole%) was added into the mixture. This was followed by dropwise addition of *m*-isopropenyl-α,α-dimethylbenzyl-isocyanate (*m*-TMI, 1.65 g, 8.21 mmole) which was diluted in DMF (10 ml) over a period of 15 minutes. The mixture was heated at 70°C for 8 hours. After cooling, the resultant solution was poured into the excess amount of cold water (1 liter). Precipitated red solid was collected and dried under vacuum at 70°C for 24 hours. It was dissolved into chloroform and the solution was dried over sodium sulfate. After concentrating the final solution, it was recrystallized into ethanol. Percent yield, 70%(wt.) m.p. 120–121°C. ^{1}H-NMR (DMSO-d^6): δ 8.19 (d, 2H), 7.78 (d, 2H), 7.40 (s, 1H), 7.27 (s, 1H), 7.45 (d, 2H), 6.95 (d, 2H), 4.01 (s, 2H), 3.35 (s, 4H), 2.08 (9H), 1.54 (s, 6H), 5.06, 5.36 (s, 2H), 7.55 (s, 1H), 7.37 (s, 3H). Anal. Calcd. for $C_{33}H_{38}N_4O_4$ (554.69): C, 71.46; H, 6.91; N, 10.10. Found: C, 71.10; H, 6.10; N, 9.95.

Poly[1-[4-(N-ethylenepiperazyl)-4'-nitrostilbene]-N-[(1,1- dimethyl-*m*-isopropenyl)benzyl]]carbamate-co-methylmethacrylate: CP-MMA (IV)

[1-[4-(N-ethylenepiperazyl)-4'-nitrostilbene]-N-[(1,1-dimethyl-*m*-isopropenyl)benzyl]] carbamate (III) (1 g, 2.26 mmole) was dissolved in 8.9 ml of freshly dried dimethylform-amide in a vacuum ampoule. To the solution, are added 0.67 g of methylmeth-acrylate and 5 mole percent of azobisisobutyronitrile. The mixture was degassed by standard freeze-vacuum-thaw technique and heated in a sealed vacuum ampoule at 60°C for 64 hours. Then the resultant mixture is poured into methanol to precipitate the copolymer. The copolymer was purified by reprecipitation from chloroform into methanol and was dried *in vacuo* at 100°C for 48 hours.

Poly{*p*-hydroxyphenylmaleimide-co-[1-[4-(N-ethylenepiperazyl)-4'-nitrostilbene]-N-[(1,1-dimethyl-*m*-isopropenyl) benzyl]]carbamate}: CP-HPMI (VI)

661 mg of HPMI (3.50 mmole) and 3.80 g of NLO monomer (7.00 mmole) were introduced into the vacuum ampoule. Freshly distilled dimethylformamide (38 ml) and azobisisobutyronitrile (AIBN, 17.1 mg) were added to the ampoule. The solution was degassed. The polymerization was carried out at 80°C for 24 hours. Then, the resulting solution was poured into hot methanol to precipitate the copolymer. The copolymer was purified by reprecipitation from tetrahydrofuran into ethyl ether and dried *in vacuo* at 100°C for 48 hours.

Poly{glycidyl methacrylate-co-[(methacryloxyethyl) methylamino]-4-nitrostilbene}: CP-GMA (VIII)

0.81 g (5.7 mmole) of glycidyl methacrylate (GMA) and 2.1 g (5.7 mmole) of 4'-[(methacryloxyethyl) methylamino]-4-nitrostilbene were dissolved in 25 ml of dimethylformamide (DMF) and put into a polymerization ampoule. 1 mol% of azobisisobutyronitrile (AIBN) was also added. Then the polymerization was carried out at 80°C for 72 hours. The resulting solution was cooled and poured into ethyl ether. The precipitate was filtered and purified by reprecipitation from tetrahydrofuran (THF) into ethyl ether and n-hexane. The product copolymer was dried in vacuo at 40°C for 72 hours. The copolymer, P2ANS of methyl methacrylate (MMA) and 4'-[(methacryloxyethyl) methylamino]-4-nitrostilbene was also prepared according to the literature method [11] to compare the EO properties with CP-GMA.

Characterization

The chemical structure of the copolymer was identified and analyzed by FT-IR (Alpha Centauri) and NMR (Bruker AM 200) spectrometer. Absorption spectra of the copolymers were recorded with a Shimadzu UV-3101 PC spectrophotometer. Differential scanning calorimeter (DSC, Perkin-Elmer DSC 7) was used to investigate the glass transition temperature of the copolymer. Elemental analysis was performed at Advanced Analytical Lab, in KIST.

Material Processing

For thin film fabrication, the synthesized copolymers, CP-MMA, CP-HPMI and CP-GMA was dissolved in tetrahydrofuran (THF)/cyclohexanone (conc. 10 wt%). Each polymer film was spin coated at 2000 rpm either on indium tin oxide (ITO) precoated glass or normal microslide glass using filtered solution (10 wt%). When tetramethylene-m-xylidene diisocyanate (m-TMXDI) of the crosslinker was mixed into the polymer solution of CP-HPMI, the mole ratio of copolymer and crosslinker was set 4 : 1. The thickness of the film was measured using a stylus instrument Tencor P10. The refractive index of the copolymer coated on silicon wafer was measured with Rudolph Auto ELII ellipsometer (Rudolph Research Co.). The wavelength used for this measurement was 632.8 nm and the incident angle was 70°.

For SHG experiment, we poled the films on microslide glass using the corona poling technique in a wire-to-plane geometry [13,14]. For linear electro-optic coefficient measurement, we deposited the gold electrode on the top of the film and did electrode poling to apply the electric field directly to both gold and ITO electrode.

Measurement of the Second-order NLO Coefficient, d_{33}

The second harmonic generation (SHG) measurements of poled samples were carried out with a Q-switched mode locked, Nd^{+3}:YAG laser operating in the TEM_{00} mode. We followed the standard Maker fringe technique which was already well understood [15]. Assuming the Kleinman's symmetry rule, we used the p–p fringe to calculate the d_{33} value. The second harmonic signal was normalized with respect to that from a calibrated quartz crystal for which value of $d_{11} = 0.5$ pm/V was assumed [16].

Measurement of Electro-optic Coefficient, r_{33}

We measured the linear electro-optic coefficients by way of simple reflection technique proposed by Teng *et al.* [17]. The He–Ne laser (wavelength: 632.8 nm) was used for this measurement. In this measurement, the sine wave modulating voltage (10 V at 8 kHz) was applied to each sample. The linear electro-optic coefficient. "r_{33}" could be calculated by following equation:

$$r_{33} = \frac{3\lambda I_m}{4\pi V_m I_c n^2} \frac{(n^2 - \sin^2\theta)^{1/2}}{\sin^2\theta}, \tag{1}$$

where I_m is the amplitude of modulation, V_m is the modulating voltage applied to the sample, and I_c is half the maximum intensity of modulation. The value determined will correspond to r_{33} provided that the electro-optic coefficient is dispersionless and the poled polymer belongs to the point group (α, ∞, ∞).

RESULTS AND DISCUSSION

1. Heterogeneous Crosslinkable Polymer: Poly{p-hydroxyphenylmaleimide-co-[1-[4-(N-ethylenepiperazyl)-4'-nitrostilbene]-N-[(1,1-dimethyl-m-isopropenyl)benzyl]]carbamate} (VI)

{1-[4-(N-ethylenepiperazyl)-4'-nitrostilbene]-*N*-[(1,1-dimethyl-isopropenyl)benzyl]}-carbamate was synthesized by a reaction of 4-[N-(2-hydroxyethyl)piperazyl]-4'-nitrostilbene with *m*-isopropenyl-α,α-dimethylbenzyl-isocyanate (*m*-TMI) in the presence of catalyst. High reaction yield could be achieved around 70%.

As shown in Scheme 1, the NLO monomer (III) was polymerized with methyl-methacrylate (MMA) to prepare the copolymer IV that is poly[1-[4-(N-ethylene-piperazyl)-4'-nitrostilbene]-N-[(1,1-dimethyl-*m*-isopropenyl) benzyl]] carbamate-co-methylmethacrylate: CP-MMA (IV). The yield was 55–60% and the inherent viscosity was measured about 0.25 dl/g in DMF at 25°C. Moreover, The NLO monomer (III) was copolymerized with HPMI through free radical polymerization as depicted in Scheme 2.

The yield was 44–45% and inherent viscosity was 0.32 dl/g in DMF at 25°C. The copolymers was quite soluble in common organic solvents such as DMF, THF, and cyclohexanone. The resultant mole ratios were calculated by integration of ¹H-NMR chemical shift which characterized each component. With the feeding ratio of 1:1 (MMA:NLO) and 1:2 (HPMI:NLO monomer), the resultant mole ratios of each component in the copolymer was determined 2:1 and 1:0.95, respectively.

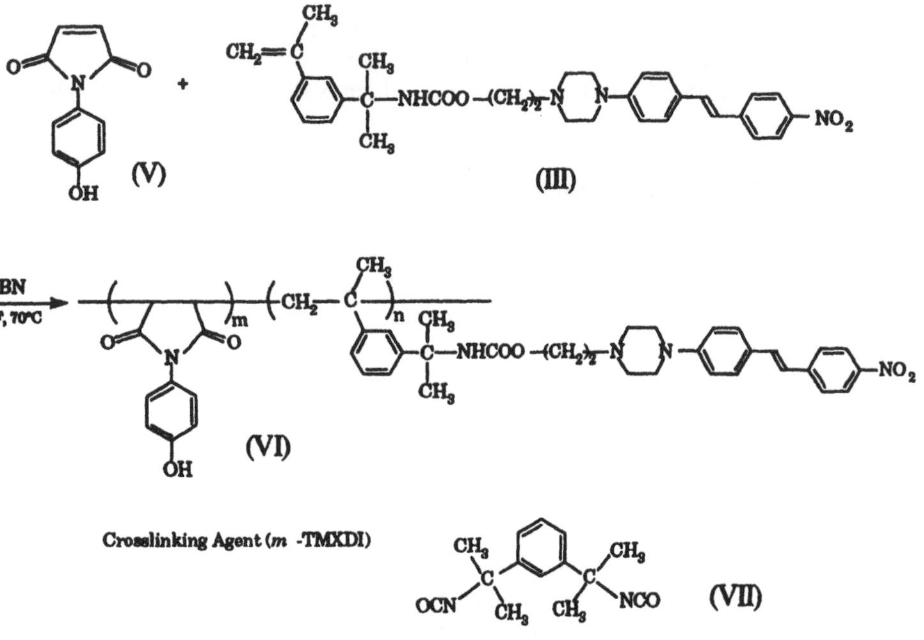

Scheme 2 Synthetic procedure of CP-HPMI and used crosslinking agent.

DSC was used to investigate the thermal transition behavior of the copolymer. DSC thermogram showed that this copolymer was amorphous judging from the absence of the melting transition and a glass transition temperatures appeared at 120–125°C of CP-MMA and 201–203°C of CP-HPMI due to the robust maleimide structure. The glass transition temperature can be controlled with the feeding amount of MMA or HPMI.

Second-order NLO Properties of CP-MMA, CP-HPMI

The simplest NLO copolymer, CP-MMA was dissolved in many common solvent to form a film. The film was poled at 120°C, which showed the tendency of molecular orientation to the poling direction, estimated by UV-VIS spectral analysis. The initial r_{33} value was calculated around 20 pm/V at 632 nm wavelength. We also fabricated the thin film of CP-HPMI and poled this film at 180°C with the change of poling field. The EO coefficient, r_{33} increased linearly with the electric field (see Figure 1). Measured r_{33} value was in the range of 25–30 pm/V at 632.8 nm wavelength when the film was poled under 0.8–1 MV/cm at 180°C. SHG measurement was performed to calculate the second-order NLO coefficient, d_{33} value of the poled film. d_{33} value was determined 14 pm/V after corona poling under 5 kV at 180°C. All measured parameters of the copolymers were tabulated in Table 1.

The temporal stability of NLO activity was investigated monitoring the decay of NLO coefficient as a function of time at elevated temperatures. After the samples

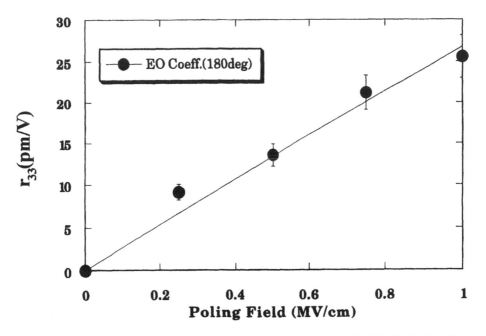

Figure 1 Poling field dependence of r_{33} of maleimide copolymer(CP-HPMI). Poling Temperature 180°C.

of CP-MMA, CP-HPMI, and crosslinked CP-HPMI being subjected to thermal aging at 80°C for 40 hours, significant difference can be observed from Figure 2. In the case of CP-MMA, the extent and rate of decay appeared considerably higher than the other two. Even after 48 hours, the value was observed to keep decreasing. However, in CP-HPMI and crosslinked CP-HPMI, a reduction of only 20% and 5% in the r_{33} value was observed for the poled films, respectively. Thermal relaxation behavior of r_{33} can give us important information for practical NLO applications. When we select the behavior of r_{33} of CP-HPMI, we can take into account as follows. After exponential decay of r_{33} at first stage of relaxation, the r_{33} value did not vary at the level of 80% of initial r_{33} value.

The decaying curve of polymer shows two different stages; at first stage, EO coefficient rapidly decreased and then the value slowly decreased approaching to a certain value ($\sim$80%) asymptotically at second stage (see Figure 2). Therefore, the electro-optic effect of this copolymer was considered quite outstanding resulting from the fairly high EO coefficient and its thermal stability. It can be thought that the second stage of relaxation should be more emphasized than the first stage for long-term application. Based on the theoretical approach, we can fit this data to a biexponential of the form, $A\exp(-t/\tau_1)+(1-A)\exp(-t/\tau_2)$. For this case, $A = 0.156$, $\tau_1 = 159$ minutes, and $\tau_2 = 120,000$ minutes. The experimental data cannot be fit to Kohlrausch–Williams–Watts stretched exponential function, $\exp\{(-t/\tau)^\beta\}$. Therefore, the r_{33} value of this polymer showed the fast decay initially and then decreased very slowly.

Table 1   Measured and calculated parameters of CP-HPMI and CP-GMA polymers.

	Absorption Maximum λ_{max} (nm)	Cut-off Wavelength (nm)	Inherent[1] Viscosity (dl/g)	T_g (°C)	Refractive[2] Index	r_{33} (pm/V)[3]	d_{33} (pm/V)[4]
CP-HPMI with TMXDI	430	605	0.32	202	1.685	30.0	14.0
CP-GMA	435	590	0.16	119	1.680	57.0	45.5

[1] Measured in dimethylformamide at 25°C.
[2] Determined by ellipsometry measurement at 632 nm.
[3] EO coefficients (at 632 nm).
[4] Second-order NLO coefficients (at 1064 nm).

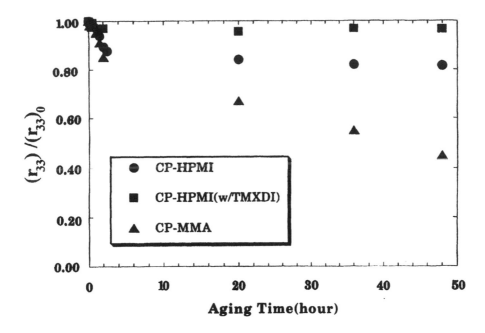

Figure 2 Decaying behaviors of r_{33} in poled copolymers films with the temperature maintained at 80°C.

Copolymer containing hydroxyl group in the side chain, which can be crosslink-able with diisocyanate inter- or intra-molecularly. Additionally, the NLO monomer used herein contains the carbamate group in the side chain. There is a secondary amine which can react with isocyanate to form biuret or allophanate. It was found that our copolymer with p-hydroxyphenyl maleimide could be thermally crosslinked in an atmosphere without deterioration of NLO activity around 140–150°C. The extent of crosslinking will be dependent on curing time and temperature. During poling, the crosslinking will be induced at the same period. This poling procedure also has a sequential temperature program to pole and then cure the film. We adopted two temperatures, 130°C and 160°C. First, we poled the film at 130°C for 1 hour and then raised the temperature to 160°C gradually in the presence of electric field. After keeping that condition for 1 hour, the poling was completed.

Using corona poling method, the UV-VIS spectra of unpoled and poled film of crosslinked polymer were compared. In Figure 3, after two-stage poling, the absorption intensity decreased and the absorption peak shifted slightly toward shorter wavelength. Then, the films were aged at 150°C for 1 hour. Usually, the absorption intensity increases due to randomization of dipolar molecules when aged at high temperature. The film with crosslinking agent showed no change of absorption intensity. Additionally, the solubility of the film was totally different from that before poling. We can believe that the dipolar alignment to poling direction was completely locked up and the stability was quite good upto 150°C. Additionally, it was thought that the copolymer system did not degrade or vaporize throughout the whole period of aging.

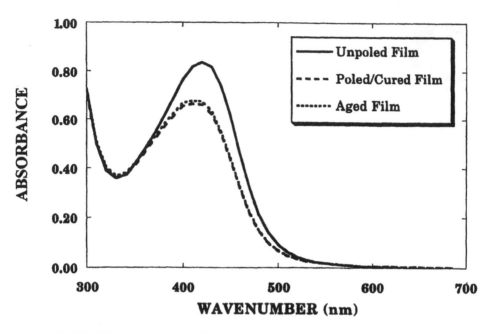

Figure 3 UV-VIS spectra of unpoled, poled, and aged film of the crosslinked CP-HPMI.

The FT-IR spectra showed well that the crosslinking reaction between the iso-cyanate and hydroxyl group in the HPMI occurred. We recorded the FT-IR spectra before curing and after curing at 110°C. The IR peaks at 2270 cm^{-1} from isocyanate and 3500 cm^{-1} from hydroxyl group decreased after curing (see Figure 4). This indicates the partial crosslinking reaction occurred through the isocyanate group and the hydroxyl group to form carbamate linkage between the side chains in p-hydroxyphenyl maleimide.

Similar to the copolymer film without crosslinker, the EO coefficient, r_{33} of the crosslinked polymer was observed around 25–30 pm/V, which depends on the poling field. In an attempt to investigate the temporal stabilities of the films with cross-linker, we aged the sample at elevated temperatures. Since the temporal stability of NLO activity should be considered to extend the period for device application, we performed the accelerating aging test at various high temperatures. Aging tempera-tures were selected 80°C, 100°C, 120°C, and 150°C. We traced the decay of r_{33} over 48 hours at each temperature. The pure copolymer film showed the 80% of the initial value of r_{33} after 2 day aging at 80°C whereas the crosslinked film maintains its r_{33} upto 98% of initial signal over the same period. The crosslinked copolymer showed quite good temporal stability at 80°C compared to the copolymer itself. The crosslinked samples aged at 100°C, and 120°C for 48 hours showed 95% and 90% of the initial EO signal, respectively. Even after 48 hours at 150°C, 80% of the residual signal was remained (see Figure 5).

We finally compared our r_{33}, d_{33} and temporal stability of our crosslinked copolymer with those of known NLO polymers. The absolute second-order NLO

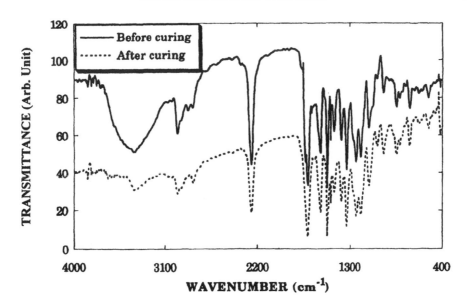

Figure 4   IR spectra of CP-HPMI with crosslinking agent. A: Before curing, B: After curing.

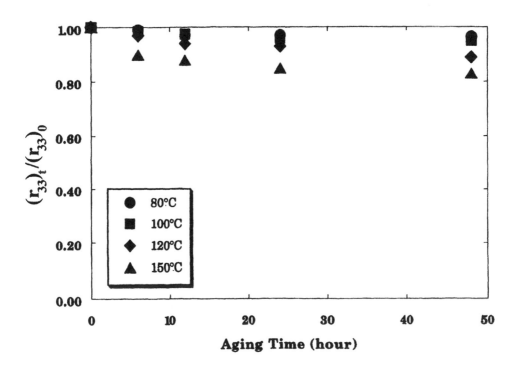

Figure 5   Decay of r_{33} of the crosslinked CP-HPMI under various temperatures.

coefficient is relatively high, however the value can be varied under the different poling condition. The temporal stability itself at a high temperature was observed better than the others in our crosslinked copolymer [18–20].

We observed good thermal stability of the second-order NLO properties upto 150°C using the highly crosslinked maleimide copolymer without degradation of the NLO chromophores.

2. Homogeneous Crosslinkable Polymer: Poly{glycidyl methacrylate-co-[(methacryloxyethyl) methylamino]-4-nitrostilbene}: CP-GMA (VIII)

An NLO monomer, 4'-[(methacryloxyethyl) methylamino]-4-nitrostilbene, was synthesized reacting 4'-[(hydroxyethyl)methylamino]-4-nitrostilbene with methacryloyl chloride. The structures of our GMA copolymer (CP-GMA) and P2ANS (IX) are shown in Scheme 3.

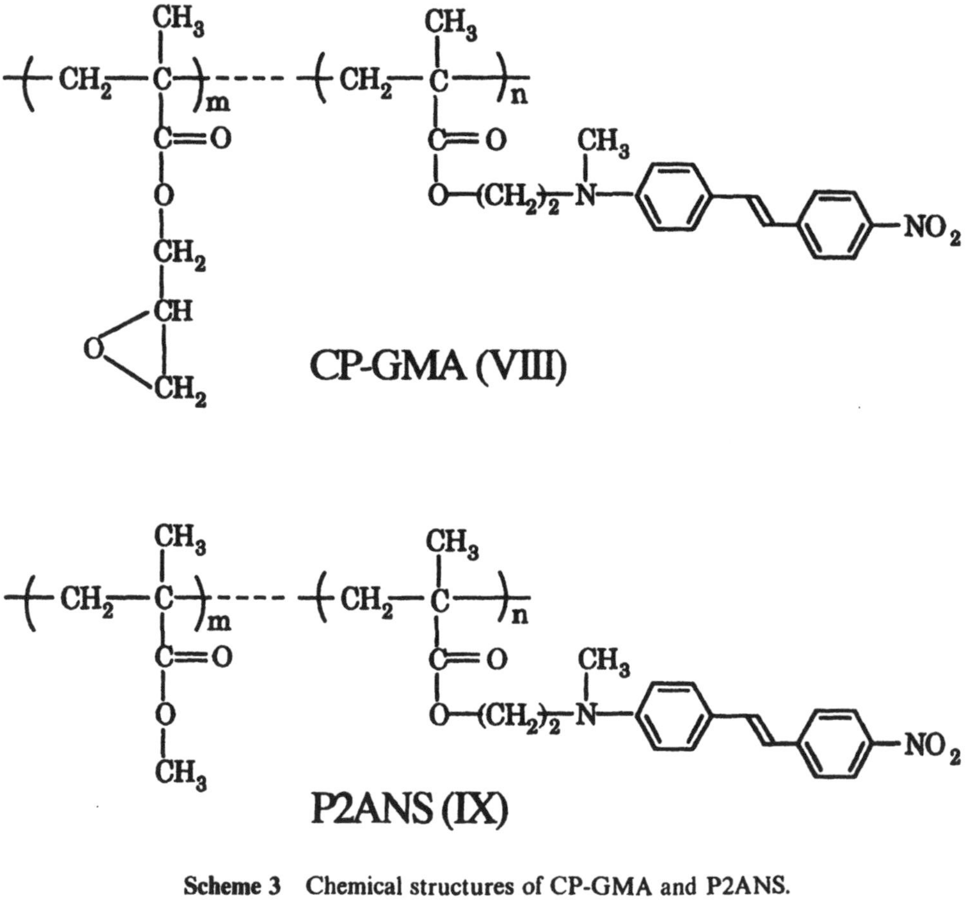

Scheme 3   Chemical structures of CP-GMA and P2ANS.

The copolymer exhibited good solubility in DMF, THF, and chloroform. We derived resultant mole ratio using the Kjeldahl's microanalysis method [21]. 36 mol% of NLO chromophore was introduced to the copolymer.

The poling and crosslinking procedures should be performed through two steps and the T_g of the copolymer was observed at 118–120°C. We, therefore, poled the polymer sample at 120°C and raised the temperature gradually for proceeding the self-crosslinking during poling (see Figure 6). UV-VIS spectra of poled and unpoled films of the copolymer were compared to investigate the extent of molecular orientation along the direction of the poling electric field. The absorption maximum of the copolymer appeared around 430–440 nm. The absorption spectrum of poled film exhibited a decrease in the absorption intensity resulting from the molecular dipolar arrangement in the direction of the poling field. Table 1 summarizes all measured and calculated parameters of the copolymer. The r_{33} value of the poled films maintained at 100°C is shown in Figure 7. The r_{33} values obtained for the respective films have been normalized to the initial r_{33} value. A comparison of the two curves shown in Figure 7 indicates that the CP-GMA has better long-term thermal stability at 100°C than P2ANS does. Thus it is thought that a certain reaction might be performed at the epoxy groups in the GMA during the poling process. Recently GMA has been used as a comonomer in polymer deriving the crosslinked polymer structure through its reactivity toward carboxylic acid attached onto the NLQ chromophore [22]. But the EO coefficient is somewhat low because the crosslink reaction hinders the molecular orientation. We confirmed the crosslink reaction between the epoxy groups

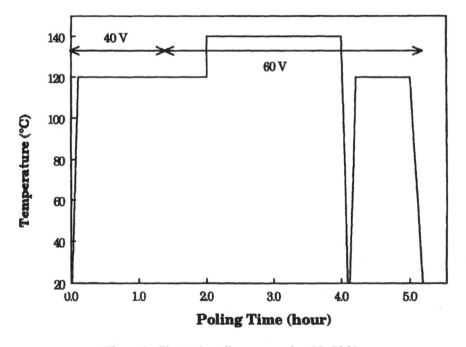

Figure 6   Electrode poling process for CP-GMA.

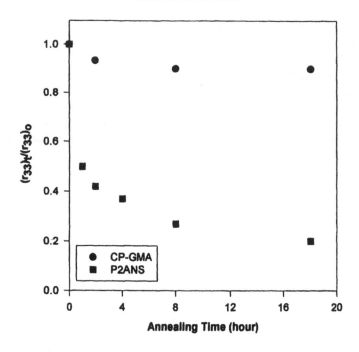

Figure 7   Decay of r_{33} of CP-GMA and P2ANS. (Aging temperature 100°C).

in the copolymer under poling. Compared to the unpoled sample, poled sample was not dissolved in any solvent. Additionally, IR spectroscopy told us that epoxy group reacted to form crosslink. The sample was treated at 140°C via thermal aging, taken out at convenient intervals, and the intensity of the epoxy characteristic band at $907\,cm^{-1}$ was measured in IR spectroscopy. The intensity of the epoxy group at $907\,cm^{-1}$ was observed to decrease with aging time, which means decrease of the epoxy group in the reaction system. The carbonyl stretching band at $1730\,cm^{-1}$ was used as a standard for this measurement because its intensity should not change during the reaction. Indeed, IR spectra of the poled films shows that the fraction of epoxy unreacted after poling was remained about 75% (Figure 8). We also investigated the absorbancy ratio change in pure poly(glycidyl methacrylate) (PGMA) film at 100°C with the reaction times. The intensity of the epoxy groups in copolymer did not decrease at a prolonged reaction time. This means that the chemical reaction between epoxy groups in the homopolymer has not occurred. Tanaka [23] reported that self-crosslinkable polyepoxides could be obtained from copolymerization of 2,3-epoxy-1-propyl methacrylate and vinyl pyridine by means of catalytic activity of tertiary amine (pyridine group in vinyl pyridine). Many literature have revealed that the organic bases, e.g. tertiary amines are Lewis bases suitable for curing epoxy resins [24,25]. We have epoxy group in GMA and tertiary amine in NLO monomer. Thus it is thought that spontaneous self-crosslinking of our NLO copolymer occurred between the epoxy groups in the presence of tertiary amine during poling even in the solid state.

2D Graph 2

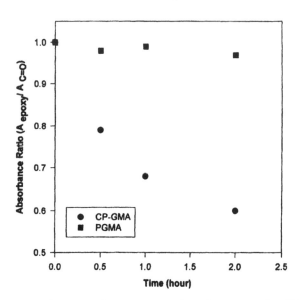

Figure 8 Fraction of unreacted epoxy group in polymer matrix. CP-GMA was poled and cured at 140°C; PGMA.

In short, studying the second-order NLO properties of two different crosslinking materials, significant improvement of the temporal stability was achieved at high temperature. Besides the crosslink between hydroxy groups via diisocyanate, particularly, the self-crosslinking mechanism between the epoxy group and tertiary amine group was proposed as a new strategy. We observed good thermal stability of the second-order NLO properties using the partially crosslinked copolymers without much degradation of the NLO chromophores.

REFERENCES

1. P. N. Prasad and D. J. Williams, in *Introduction to Nonlinear Optical Effects in Molecules and Polymers*, Wiley, NewYork, 1991.
2. D. J. Williams, in *Nonlinear Optical Properties of Organic and Polymeric Materials*, ACS Symp. Ser. 233, 1983.
3. D. S. Chemla and J. Zyss, in *Nonlinear Optical Properties of Organic Molecules and Crystals*, Academic Press, Orlando, FL, 1987.
4. A. J. Heeger, J. Orenstein and D. R. Ulrich, in *Nonlinear Optical Properties of Polymers*, Mater. Res. Soc. Symp. Proceedings, Pittsburgh, 1987, Vol. 109.
5. P. N. Prasad and D. R. Ulrich, in *Nonlinear Optical and Electroactive Polymers*; Prasad, Plenum Press, NY, 1988.
6. R. A. Hann and D. Bloor, in *Organic Molecules for Nonlinear Optics*, The Royal Society of Chemistry Publication, London, 1989, Vol. 69.

7. J. Messier, F. Kajzar, P. N. Prasad and D. R. Ulrich, in *Nonlinear Optical Effects in Organic Polymers*, NATO ASI series, Kluwer Academic Publishers, Netherlands, 1988, Vol. 162.

8. C. Ye, T. J. Mark, J. Yang and G. K. Wong, *Macromolecules*, **20**, 2322 (1987).

9. D. H. Choi, H. M. Kim, W. P. K. M. Wijekoon and P. N. Prasad, *Chem. Mat.*, **4**(6), 1253 (1992).

10. R. N. Martino (Hoechst Celanese Corp.), *Eur. Pat.* 0 294 706 (1988) (Chem. Abstr., 110, 16. 1989).

11. J. O. Park and S. H. Jang, *J. Appl. Polym. Sci.*, **30**, 723 (1992).

12. R. N. Martino (Hoechst Celanese Corp.), USP 4, 808, 332 (1989).

13. M. A. Mortazavi, A. Knoesen and S. J. Kowel, *J. Opt. Soc. Am. B*, **6**, 733 (1989).

14. W. M. K. P. Wijekoon, Y. Zhang, S. P. Karna, P. N. Prasad, A. C. Griffin and A. M. Bhatti, *J. Opt. Soc. Am. B*, **9**, 1832 (1992).

15. K. D. Singer, J. E. Sohn and S. J. Lalama, *Appl. Phys. Lett.*, **49**, 248 (1986).

16. J. Jerphagnon and S. K. Kurtz, *Phys. Rev. B*, **1**, 1739 (1970).

17. C. C. Teng and H. T. Mann, *Appl. Phys. Lett.*, **56**(18), 30 (1990).

18. R. J. Gulloty, D. J. Brennan, M. A. Chartier, J. K. Gille, A. P. Haag, K. A. Hazard and M. N. Inbasekaran, in *Organic Thin Films for Photonic Applications*, ACS/OSA, Optical Society of America, Washington, D.C. 1993, p. 6.

19. P. M. Rannon, Y. Shi, W. H. Steier, C. Xu, B. Wu and L. R. Dalton, in *Organic Thin Films for Photonic Applications*, ACS/OSA, Optical Society of America, Washington, D.C. 1993, p. 10.

20. W. M. Herman, L. M. Hayden, S. Brower, G. A. Lindsay, J. D. Stenger-Smith and R. A. Henry, in *Organic Thin Films for Photonic Applications*, ACS/OSA, Optical Society of America, Washington, D.C. 1993, p. 18.

21. G. F. Jeffery, J. Basett, J. Mendham and R. C. Denney, Textbook of Quantitative Chemical Analysis, 5th ed. (Longman Scientific & Technical, 1989) p. 302.

22. R. Levenson, J. Liang and J. Zyss, *Polym. Prepr.*, **35**, 162 (1994).

23. R. S. Bauer, ACS Symposium Series No. 114, Epoxy Resin Chemistry, Vol. I, 1979, p. 197.

24. H. Lee and K. Neville, Handbook of Epoxy Resins, Chap. 5, 1967, p. 54.

25. Kirk-Othimer, Encyclopedia of Chemical Technology, Vol. 9, 1980, p. 267.

3. NEW POLYAMIDE MAIN-CHAIN POLYMERS BASED ON 2′,5′-DIAMINO-4-(DIMETHYLAMINO)-4′-NITROSTILBENE (DDANS)

CHRISTOPH WEDER [a], BERNHARD H. GLOMM [a],
PETER NEUENSCHWANDER [a], ULRICH W. SUTER [a],
PHILIPPE PRÊTRE [b], PHILIP KAATZ [b], and PETER GÜNTER [b]

[a]*Institut für Polymere, Eidgenössische Technische Hochschule, CH-8092 Zürich, Switzerland;* [b]*Institut für Quantenelektronik, Eidgenössische Technische Hochschule, CH-8093 Zürich, Switzerland*

ABSTRACT

Polyamides based on 2′,5′-diamino-4-(dimethylamino)-4′-nitrostilbene (DDANS) represent a new approach to the design of NLO polymers in which the nonlinear optical units are fixed in the polymer backbone with their dipole moments oriented transversely to the main chain. Two different series of polymers have been synthesized and investigated. Semiflexible polyamides based on DDANS and linear aliphatic diacid chlorides are amorphous and show glass transition temperatures up to 206°C. They can easily be processed by spin-coating into thin films of optical quality. Second-order nonlinear optical coefficients (d_{33}) of up to 40 pm/V at $\lambda = 1.54\,\mu m$ have been measured on films which have been oriented by a corona discharge poling process. Orientational relaxation experiments reveal a remarkable orientational stability for these polymers. At ambient conditions, no significant relaxation of the NLO coefficients can be observed within 240 days after poling. Rigid-rod polyamides (aramids) based on DDANS and aromatic diacid chlorides are a class of novel liquid-crystalline polymers, where different mechanisms of orientation can be combined. In the nematic phase, the rigid-rod molecules form highly aligned domains that can be oriented using mechanical processes such as shearing. In addition, the stilbene units, which are fixed in the polymer backbone with their dipole moments oriented transversely to the main chain, can be oriented in an electric field. The combination of shearing and corona-discharge poling was found to yield highly oriented films. The two orientation mechanisms seem to cause a synergistic effect, probably since each affects different levels of the polymer microstructure in the solid.

SEMIFLEXIBLE POLYAMIDES

In recent years there has been considerable interest in nonlinear optical (NLO) and electro-optic (EO) polymers because of their potential application in efficient, ultrafast and low-voltage integrated electro-optical devices [1]. The frequently cited advantages of poled polymers are large nonlinear susceptibilities, fast response times, and easy processability. However, additional requirements such as long-term stability of the NLO or EO effect, high physical and mechanical stability and low optical propagation losses must be satisfied before polymeric materials can demonstrate their potential as actual device materials [1].

Insufficient temporal stability of the induced noncentrosymmetric orientation of the NLO-phores at elevated temperatures, as they are expected for device processing (up to 250°C) and operation (80°C), is a major hindrance to further progress in developing polymeric NLO devices [2]. Different design strategies have been proposed for polymers with enhanced orientational stability of the NLO-phores. Generally, polymer systems with increased glass transition temperature show an improved orientational stability of the NLO-phores, although the efficacy varies depending on how the NLO-phore is incorporated into the polymer matrix. Covalent connection of the NLO-phore to the polymer is advantageous; typically the NLO-phores have been attached as side-chain to the polymer backbone [3]. Other approaches are based on cross-linked polymers [4] or polymers which incorporate the NLO-phores with their dipole moments head-to-tail in the main chain [5]. Recently we proposed a new approach to NLO polymers with large and stable second-order nonlinear susceptibilities where the NLO-phores are part of the polymer backbone and are linked with small spacer units [6]. The dipole moments of the NLO-phores are oriented transversely to the polymer main chain, a consequence of the assumption that in this arrangement the NLO-phores are easier to orient by an external field than in structures where their dipole moments are pointing along the polymer main chain (see Figure 1).

Structure and Chemical Properties

We have synthesized a series of semiflexible polyamides based on the bifunctional NLO-phore 2′,5′-diamino-4-(dimethylamino)-4′-nitrostilbene (DDANS) and different linear aliphatic diacid chlorides as shown in Figure 1 [6]. A low-temperature polycondensation reaction was used in order to obtain polymers with weight-average molecular weights (M_w) of up to 116 000. The polymers are amorphous, showing glass transition temperatures (T_g) that linearly depend on the number of methylene groups of the diacid chloride and range from 125°C to 206°C. TGA and DSC experiments show that the most significant decomposition step for all polymers sets in at around 210°C in air and also under nitrogen atmosphere. However, time- and temperature-resolved UV-vis analysis of the NLO-phore's charge-transfer

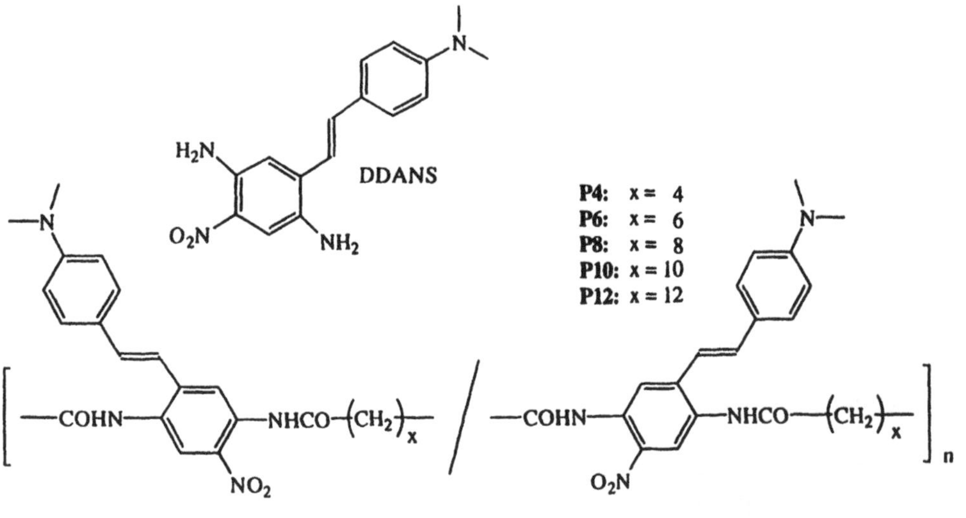

Figure 1 New semiflexible NLO main-chain polyamides based on DDANS.

absorption band revealed, that chemical degradation of the NLO-phores is an important process even at temperatures well below the onset temperature determined by DSC and TGA. All polymers are completely soluble in concentrations of 5% w/w or more in classical amide solvents such as 1-methyl-2-pyrrolidone (NMP) and dimethyl sulfoxide (DMSO) and acids such as formic acid or trichloro acetic acid. Thin films of good optical quality can easily be obtained by common spin coating techniques (see Table 1).

Linear and Nonlinear Optical Properties

Polymers **P12** and **P6** were our main points of interest because these are the polymers with the lowest and the highest glass transition temperatures of the prepared series (omitting **P4**, the T_g of which is 206°C, in the range of thermal decomposition). Therefore, the linear and nonlinear optical properties of these two polyamides were examined in greatest detail. All polymers show an absorption maximum between 471 and 473 nm and are transparent at wavelengths longer than 700 nm. The extinction coefficients are slightly different due to the different contents of NLO-phore of the polymers.

Transparent thin films of **P6** and **P12** were oriented by a corona-discharge poling process. The nonlinear optical coefficients d_{33} have been determined using a standard Maker-fringe technique [7] at a fundamental wavelength of 1542 nm. The absorption of the polymers at the second harmonic wavelength (771 nm) is negligible and, therefore, the obtained NLO coefficients d_{33}, 40 pm/V for **P12** and 27 pm/V for **P6**, are essentially not resonance enhanced. The electronic contribution r_{33}^e to the electro-optic coefficients r_{33} has been calculated to be 16 pm/V (**P12**) and 10 pm/V (**P6**) at 1300 nm according to a direct relation between electronic electro-optic and nonlinear optical coefficients [8]:

$$r_{ijk}^e = \frac{-4}{n_i^2 n_j^2} \cdot d_{ijk}^{EO},$$

Table 1   Physical properties of polymers **P4–P12**.

Polymer	Chromophore concentration[a]	η_{inh},[b] dL/g	T_g, °C
P12	67.7%	0.40	125
P10	71.5%	0.73	144
P8	75.8%	0.46	159
P6	80.7%	0.82	176
P4	86.3%	0.89	206

[a] In % w/w of the polymer's repeat unit.
[b] Inherent viscosity at 25°C in NMP at a polymer concentration of 0.5 g/dL.

where the nonlinear optical coefficient d_{ijk}^{EO} for a static and an optical field at frequency ω is related to the nonlinear optical coefficient $d_{kij}^{-2\omega',\omega',\omega'}$ measured at optical frequencies ω' by

$$d_{ijk}^{EO} \equiv d_{ijk}^{-\omega,\omega,0} = \frac{f_i^\omega f_j^\omega f_k^0}{f_k^{2\omega'} f_i^{\omega'} f_j^{\omega'}} \cdot \frac{(3\omega_0^2 - \omega^2) \cdot (\omega_0^2 - \omega'^2)^2 \cdot (\omega_0^2 - 4\omega'^2)}{(3\omega_0^2 - 3\omega'^2) \cdot (\omega_0^2 - \omega^2)^2 \cdot \omega_0^2} \cdot d_{kij}^{-2\omega';\omega',\omega'}.$$

The f are local field factors in the Lorentz approximation and ω_0 is the resonance frequency for the dominant oscillator. The nonlinear optical properties of **P6** and **P12** are summarized in Table 2.

Orientational Relaxation

The orientational relaxation of the NLO-phores in corona poled films of the semiflexible polyamides was investigated at different temperatures below the glass transition by the decay of the nonlinear optical susceptibilities [9]. The time dependence of the decay was found to be well represented by the Kohlrausch–Williams–Watts (KWW) stretched-exponential function, which is often used to describe the orientational relaxation in poled polymers [3,9]:

$$d(t) = d_0 \exp[-(t/\tau)^\beta].$$

Here τ is the characteristic relaxation time that is required for the NLO susceptibility $d(t)$ to decay to $1/e$ of its initial value d_0. The function represents in its original sense a continuous distribution of single exponential decays where $0 < \beta \le 1$ is a measure of the breadth of the distribution and the extent of deviation from single exponential behavior. We emphasize that in the present study the decay of the nonlinear optical susceptibilities cannot be attributed to a single mechanism. Time- and temperature-resolved UV-vis analysis of the charge-transfer absorption band revealed, that chemical degradation of the NLO-phores at elevated temperatures is an important process in addition to the orientational relaxation [9]. However, the KWW function was found suitable to describe the decay of the nonlinear optical susceptibilities, especially in the longer time portion of the decay. Figure 2 shows the measured decay of d_{33} for polymer **P6** and the calculated stretched exponential functions at four different decay temperatures. The results are summarized in Table 3. No significant

Table 2 Nonlinear optical properties of polymers **P6** and **P12**.

Polymer	λ_{max}, nm	d_{33}, pm/V (1542 nm)[a]	r_{33}^e, pm/V (1300 nm)[b]
P12	471	40	16
P6	473	27	10

[a] Second-order NLO coefficient, experimental error: 10%.
[b] Electronic electro-optic coefficient.

Table 3 Results of the relaxation experiments at different temperatures: Parameters of the KWW function fitted to each relaxation experiment.

	Polymer **P12**				Polymer **P6**			
	25°C	80°C	107°C	116°C	25°C	80°C	158°C	167°C
d_0, pm/V	19	23	36	34	21	25	26	23
τ, sec	$2.3 \cdot 10^{13}$	$1.7 \cdot 10^6$	$1.3 \cdot 10^4$	560	$3.2 \cdot 10^{13}$	$3.0 \cdot 10^{7}$ [a]	1140	290
β	0.15	0.27	0.31	0.47	0.20	0.55 [a]	0.27	0.42

[a] Thermal degradation processes overlap the relaxation process significantly in this case.

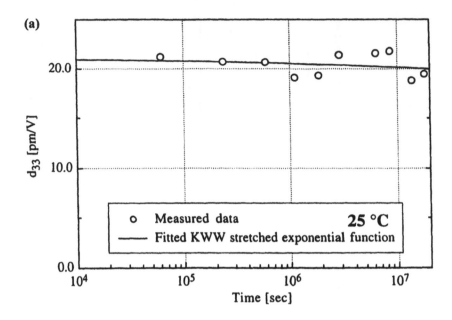

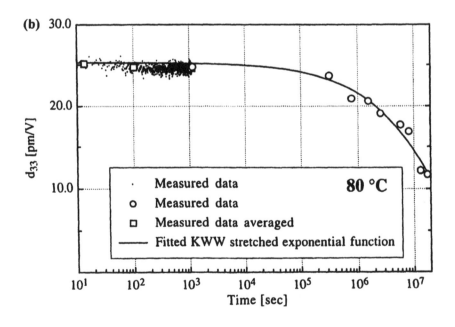

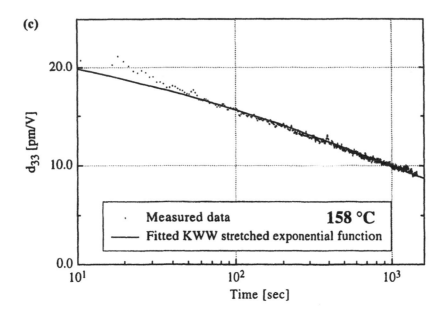

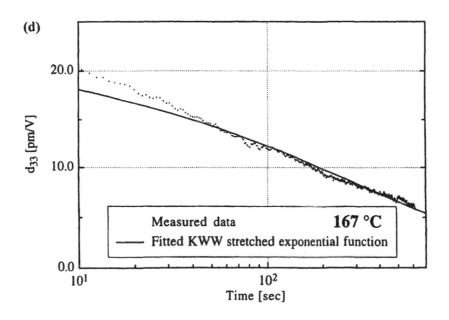

Figure 2 SHG relaxation of thin films of polymer **P6** as a function of time and temperature, determined at 25 (a), 80 (b), 158 (c), and 167 (d) °C.

decay of the nonlinear susceptibilities could be detected for polymers **P12** and **P6** at 25°C within 8 months after poling.

The temperature dependence of the decay could be correlated with the glass transition temperature T_g using a normalized relaxation law with $(T_g - T)/T$ as the relevant scaling parameter [3,9]. For the pure orientational relaxation at 80°C, a relaxation time of significantly over ten years could be estimated for polymer **P6**. We surmise that the outstanding orientational stability of the new polyamides are predominantly based on the location of the NLO-phores in the skeletal structure of the polymer backbone. This arrangement provides sufficient mobility for the orientation process but minimizes the orientational relaxation. As discussed above, the restricted thermal stability of the NLO-phore was found to be a limiting factor for the long-term behavior of the nonlinear optical susceptibility of the high-T_g polyamide investigated here. The NLO-phore used in our polymers, being based on DANS, is very similar to many commonly employed NLO-phores; hence, we suggest that the role of thermal stability of NLO-phores receives special attention.

RIGID-ROD ARAMIDS

Taking advantage of the structural characteristics of DDANS which, as discussed above, enables the incorporation of a dipolar stilbene unit into a polymer backbone, we introduced a novel class of fully aromatic and rigid-rod liquid-crystalline polyamides (aramids), where different mechanisms of orientation can be combined [10]. In the nematic phase, the rigid-rod molecules form highly aligned domains that can be oriented using mechanical processes such as shearing. In addition the highly dipolar stilbene units which are fixed in the polymer backbone with their dipole moments oriented transversely to the main chain, can be oriented in an electric field. The long-range orientational order of the macromolecules was found to be the reason for the extraordinary properties of aramids and, hence, it is an important factor in the processing of these materials to generate and maximize this order [11].

Structure and Chemical Properties

Two novel, fully aromatic rigid-rod aramids have been prepared in a low-temperature solution polycondensation from DDANS and terephthaloyl dichloride or 2,6-difluoro-terephthaloyl dichloride as shown in Figure 3. Both aramids are completely soluble in formic acid, methanesulfonic acid, sulfuric acid, NMP/LiCl, dimethyl acetamide/LiCl, and DMSO/LiCl. Polymer **P2** was also soluble in NMP without any salt addition. Liquid-crystalline solutions have been observed in all of the acids listed above. The inherent viscosity at a polymer concentration of $c = 0.5\,\text{g/dL}$ at 25°C in H_2SO_4 was 1.44 for both polymers, indicating reasonably high molecular weights of these materials. The constitutional order of these rigid-rod aramids depends strongly on the reactivity differences of the functional groups and on the detailed conditions of the polymerization [12]. Based on the actual reaction conditions and the reactivity of the used monomers [12d], we expect a constitutionally

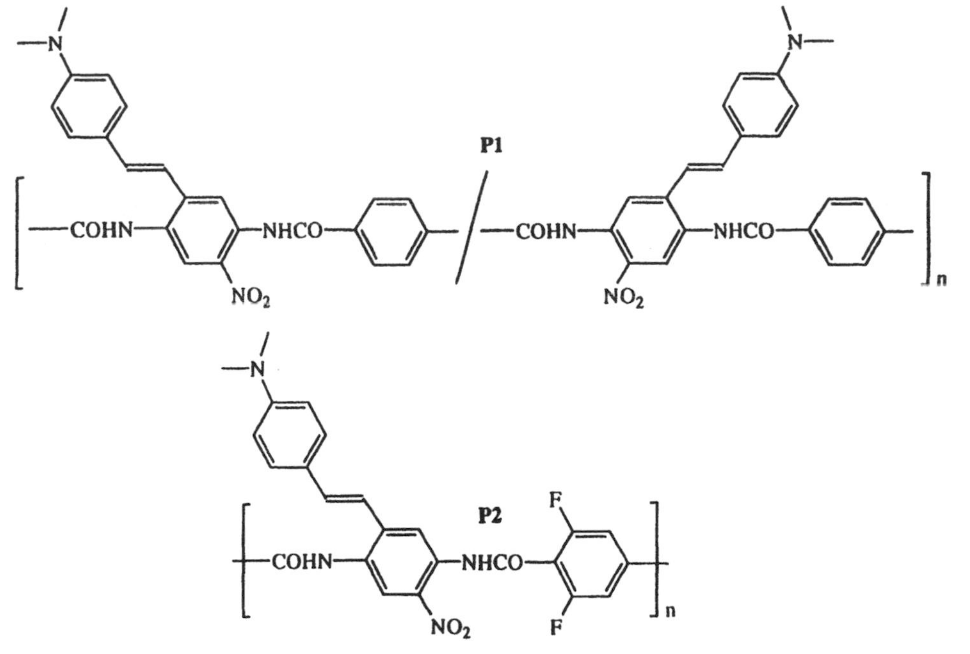

Figure 3 New rigid-rod main-chain aramids based on DDANS.

irregular arrangement of the monomeric units for **P1** and a constitutionally highly regular 'head-to-tail' ordered structure in the case of **P2** as shown in Figure 3.

Orientation and Supramolecular Structure of Films and Fibers

Nematic solutions of the constitutionally ordered polymer **P2** in methanesulfonic acid were processed into fibers and films which could be characterized by wide-angle X-ray-diffraction (WAXD) measurements [10]. To investigate the crystal structure of this polymer, fibers were prepared by a dry-jet wet spinning process [13]. A post-spin heat treatment under constant axial stress, which was applied to the fibers, gave rise to a significant improvement of chain orientation and decreased the number and size of crystal defects. Taking advantage of the uniaxially drawn fiber texture and the analogy of the polymer backbone with poly(p-phenylene terephthalamide) (PPTA) allowed for indexing the reflections in the WAXD fiber diagrams. The dominating crystal structure was indexed in analogy to a pseudo-edge-centered structure akin to an enlarged "Modification II" of the fibers from PPTA. The observed reflections could be indexed in consistence with an orthorhombic unit cell (space group Pa) with the edge lengths $a = 15.0\,\text{Å}$, $b = 14.2\,\text{Å}$, $c = 12.9\,\text{Å}$, and with one chain per unit cell. The fibers of polymer **P2** do not suffer a structural transformation upon heat treatment as it is observed for PPTA.

In order to reach an improved long-range orientational order of the macro-molecules in aramid films we combined two different ordering mechanisms as outlined above. A nematic solution of **P2** in methanesulfonic acid was pressed between two ITO coated glass slides and slightly sheared. After removing the covering slide, the films were either coagulated in a water bath, or corona poled and then coagulated. The films were subsequently washed with water for several days to remove remaining solvent and then dried under vacuum at 80°C for 2 days. The thickness of the resulting films ranged from 90 to 250 µm with roughnesses in excess of 1 µm. An X-ray study was carried out in order to measure the orientational characteristics of the films. Wide-angle X-ray diffraction patterns were collected parallel and perpendicular to the film surfaces, each for a poled and an unpoled sample. The *meridional* and *equatorial* intensity profiles of the fiber diagram for a poled and an unpoled film measured perpendicular to the surface are shown in Figure 4.

The significantly improved resolution of the most important reflections (e.g. (001), (006)) for the poled sample shows that the poling process gave rise to a remarkable gain in crystallinity and decreased the number and size of crystal defects. Further-more, the apparent orientational angle [13] ζ was about 106.5° for the unpoled sample and changed to 72° due to the poling process, which corresponds to a 1.5 fold amelioration of average chain orientation. The structural and morphological improvement for the fiber diagrams perpendicular to the sample surface is slightly higher than in the case of measurements parallel to the film surface. As described above for the fibers, no structural transformation could be observed due to the poling process. The indexing could be established exactly as done for the fiber WAXD-patterns mentioned above.

Figure 5 sketches the effects of the poling process on the supramolecular structures of the film samples. The unpoled film consists of many small frozen "nematic domains" with a broad distribution with respect to the size and the shape of these "domains". They are separated from each other by broad, more or less amorphous defect layers. The "directors" of these "domains" are nearly randomly oriented, showing only a slight preference for orientation along the vertical main sample axis. The poled film consists of comparatively few, larger, frozen "nematic domains" with a sharp distribution with respect to the size and the shape of these "domains". They are separated from each other by only a comparably small number of thin defect layers with sharp borders. The "directors" of these "domains" are preferably oriented along the vertical main sample axis.

These postulated supramolecular structures are consistent with the investigated diffraction patterns. The significant enhancement of the average polymer chain orientation in the direction of the vertical sample axis, caused by the application of the electric field perpendicular to the film surface, can be interpreted as a conse-quence of the dielectric anisotropy of the polymer backbone. Its magnitude and direction determine the orientation of a liquid crystal in an electric field. In the case of the polymer **P2** discussed here, we therefore expect a strong tendency of the large electric dipole moments of the stilbene moieties to be oriented along the electric field lines. Because of the rigidity of the fully aromatic macromolecules and the

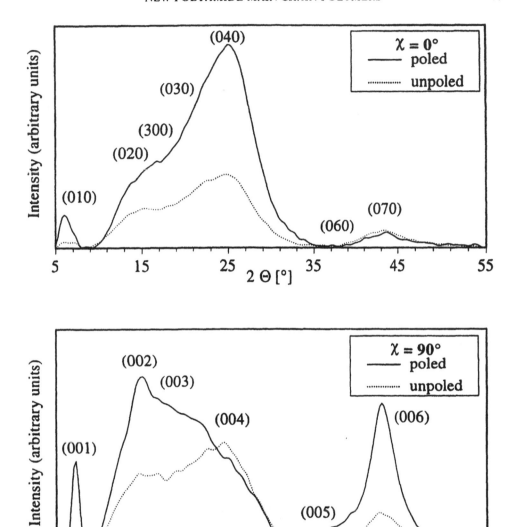

Figure 4 Equatorial (top) and meridional (bottom) WAXD scans of films of polymer P2, measured perpendicular to the film surface.

shear-induced orientation of the chain axes parallel to the film surface, an orientation of the polymer backbones perpendicular to the electric field lines is expected. Due to the noncentrosymmetric orientation of the stilbene moieties in the poled films, a second-order nonlinear optical response can be expected for these structures. However, nonlinear optical measurements on poled films of the aramids discussed

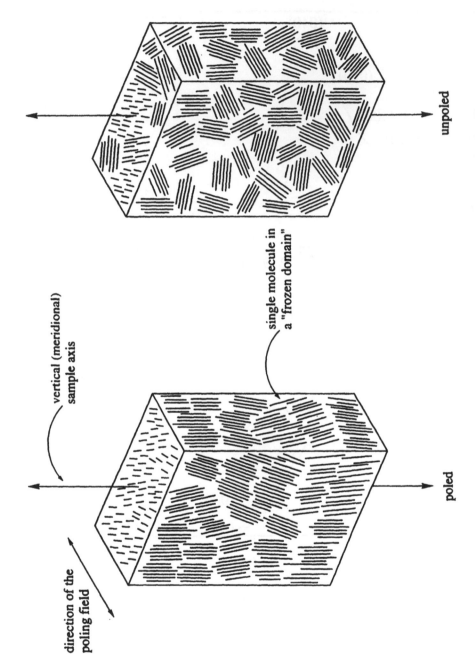

Figure 5 Schematic representation of the effects of the poling process on the supramolecular film structures.

here have been stifled so far by the insufficient optical properties of the prepared samples caused by the difficulties to process the highly viscous dope into homogeneous thin films.

CONCLUSION

We have presented a new approach to the design of NLO polymers where the NLO-phores are part of the polymer backbone and are linked with small spacer units. The new bifunctional NLO-phore DDANS was synthesized and a series of semiflexible linear polyamides containing the new NLO-phore was prepared. These polymers have a high NLO-phore density, are amorphous, and have high, adjustable glass transition temperatures. The polymers can easily be processed into thin films of optical quality. Second-harmonic generation measurements on poled films show large second-order nonlinearities. The new main-chain polyamides exhibit a remarkable orientational stability, which we relate to the location of the NLO-phores in the polymer backbone. The restricted thermal stability of the NLO-phore was found to be the limiting factor for the long-term behavior of the nonlinear optical response of the new materials.

Two novel, fully aromatic rigid-rod polyamides have been prepared in a low-temperature solution polycondensation from DDANS and terephthaloyl dichloride or 2,6-difluoro-terephthaloyl dichloride. These aramids represent a new approach to the design of lyotropic polymers, where different mechanisms of orientation can be combined, resulting in an improved long-range order in thin polymer films. In the nematic phase, the rigid-rod molecules form highly aligned domains that can be oriented using mechanical processes such as shearing. In addition, the stilbene units that are fixed in the polymer backbone with their dipole moments oriented transversely to the main chain can be oriented in an electric field. The combination of shearing and corona-discharge poling was found to yield highly oriented films. The two orientation mechanisms seem to cause a synergistic effect, probably since each affects different levels of the polymer microstructure in the solid.

ACKNOWLEDGMENT

Financial support by the *Schweizerischer Nationalfonds zur Förderung der wissenschaftlichen Forschung* (NF Sektion II) is gratefully acknowledged.

REFERENCES

[1] Burland, D. M., Miller, R. D. and Walsh, C. A. (1994). *Chem. Rev.* **94**, 31.
[2] Papers presented in *Organic Thin Films for Photonic Applications, Technical Digest Series*, Vol. 17, Optical Society of America, 1993.
[3] Prêtre, P., Kaatz, P., Bohren, A., Günter, P., Zysset, B., Ahlheim, M., Stähelin, M. and Lehr, F. (1994). *Macromolecules* **27**, 5476.

[4] Eich, M., Reck, B., Yoon, D. Y., Willson, C. G. and Bjorklund, G. C. (1989). *J. Appl. Phys.* **66**, 3241.

[5] Willand, C. S. and Williams, D. J. (1987). *Ber. Buns. Ges. Phys. Chem.* **91**, 1304.

[6] Weder, Ch., Neuenschwander, P., Suter, U. W., Prêtre, P., Kaatz, P. and Günter, P. (1994). *Macromolecules* **27**, 2181.

[7] Jerphagnon, J. and Kurtz, S. K. (1970). *J. Appl. Phys.* **41**, 1667.

[8] Bosshard, Ch., Sutter, K., Schlesser, R. and Günter, P. (1993). *J. Opt. Soc. Am. B* **10**, 867.

[9] Weder, Ch., Neuenschwander, P., Suter, U. W., Prêtre, P., Kaatz, P. and Günter, P. (1995). *Macromolecules* **28**, 2377.

[10] Weder, Ch., Glomm, B. H., Neuenschwander, P. and Suter, U. W. (1995). *Macromol. Chem. Phys.* **196**, 1113.

[11] Yang, H. H. (1990). *Aromatic High-Strength Fibers*, Wiley Interscience, New York.

[12] (a) Gentile, F. T., Meyer, W. R. and Suter, U. W. (1991). *Macromolecules* **24**, 633; (b) Meyer, W. R., Gentile, F. T. and Suter, U. W. (1991). *Macromolecules* **24**, 642; (c) Gentile, F. T. and Suter, U. W. (1991). *Makromol. Chem. Phys.* **192**, 663; (d) Grob, M. C. (1993). Ph.D. thesis Nr. 10360, ETH, Zürich.

[13] (a) U. S. Patent 3767756, **(1973)**. E.I. Dupont de Nemours and Co., inv.: H. Blades, *Chem. Abstr.* (1973) **78**, 85795w; (b) U.S. Patent 3869430 (1975). E.I. Dupont de Nemours and Co., inv.: H. Blades, *Chem. Abstr.* (1973) **78**, 31328c.

4. MULTILAYER $\chi^{(2)}$ NLO FILMS PREPARED BY THE LBK PROCESS FROM NOVEL ACCORDION POLYMERS

GEOFFREY LINDSAY[a], KENNETH WYNNE[b], WARREN HERMAN[c],
ANDREW CHAFIN[a], RICHARD HOLLINS[a], JOHN STENGER-SMITH[a],
JAMES HOOVER[a], JERROLD CLINE[c], and JOSEPH ROBERTS[a]

[a] US Navy, NAWCWPNS, Chemistry and Materials Branch, China Lake,
CA 93555, USA; [b] Office of Naval Research, Arlington, VA 22217, USA;
[c] NAWCAD, Avionics Branch, Warminster, PA 18974, USA

ABSTRACT

Polymeric thin films having second-order nonlinear optical (NLO) properties were fabricated by the Langmuir–Blodgett–Kuhn (LBK) technique. Mainchain polymers were designed and synthesized with chromophores in the syndioregic (head-to-head) configuration. Multilayer-$(AB)_n$-films were fabricated from two polymers by Y-type deposition. In Polymer A, the chromophore's electron donating end was connected to a relatively hydrophobic bridging unit, and its electron accepting end was connected to a relatively hydrophilic bridging unit. The converse was true for Polymer B. Microstructural information about these multilayer polymer films was obtained from polarized optical measurements.

INTRODUCTION

Important new chromophore-bound polymers for second-order nonlinear optical (NLO) applications, such as optical second harmonic generation and electro-optic modulation of optical signals, have recently been reported [1,2]. It is important to achieve a high degree of polar alignment of the chromophores in order to maximize the second-order NLO properties. The most often used process for achieving polar alignment is electric-field poling. In electric-field poling, the degree of polar alignment is related to $\mu E/kT$, where μ is the ground state dipole moment of the chromophore, E is the applied electric field, k is Boltzmann's constant, and T is the poling temperature. Because the polymer must be heated to near the glass transition temperature for the chromophores to have adequate mobility to be aligned, the high thermal vibrational energy of the chromophores works against the torque of the electric field.

Self-assembly processes, such as the Langmuir–Blodgett–Kuhn (LBK) process [3], can give multilayer dipole-aligned films at low temperature [4]. This paper reports the characterization of multilayer films prepared by the LBK deposition process at room temperature. The polymers designed for this study have the so-called 'accordion', or syndioregic mainchain NLO polymer topology [5], in which hydrophilic and lipophilic bridging groups are located between the chromophores in order to bring about self-ordering at an air–water interface [6].

In Polymer A, the chromophore's electron donating end is connected to a relatively hydrophobic bridging unit, and its electron accepting end is connected to a relatively hydrophilic bridging unit. The converse is true for Polymer B.

When the polymer chains are deposited at the air–water interface, it is expected that the backbones fold and position the relatively hydrophilic bridging groups near the water interface, and the relatively hydrophobic bridging groups near the air interface. Thus, the long dipole axis of the chromophores would tend to align normal to the plane of the water surface (hence, normal to the plane of the polymer film). As multiple layers are deposited by the Y-type method, in alternating (AB)$_n$ fashion, the hydrophilic surfaces of adjacent layers would be in contact, as would the hydrophobic surfaces (as this is the most thermodynamically stable conformation) [3].

Furthermore, if an inversion of dipole direction is required within the film, for example, to maximize the overlap integrals of fundamental and second harmonic for phase matching in frequency doubling applications [7], then one would need only to deposit one of the polymers twice in succession, then continue with the alternating pattern.

EXPERIMENTAL PROCEDURES AND RESULTS

Synthesis

The chemical structures of the two polymers used in this study are shown in Figure 1. The synthesis of Polymers A and B was similar to the procedure given in Ref. [5]. The detailed synthesis of Polymer B precursors are given in Ref. [8].

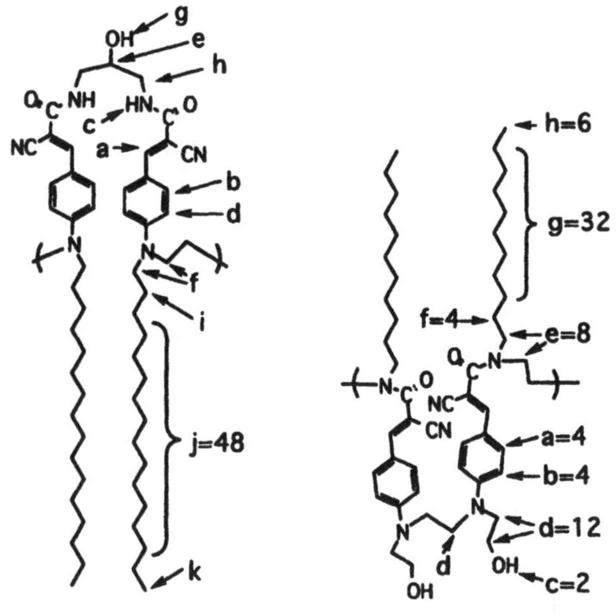

Figure 1   Letters correspond to the proton NMR assignments [9].

Polymer Characterization

The proton nuclear magnetic resonance assignments are as expected from the target polymers shown in Figure 1.

An estimate of the number average molecular weight of these polymers was made from the proton NMR spectra using the areas of the aldehyde end group resonance and the methyl group resonance. The number-average molecular weight of Polymer A was about 47000 g/mol. The number-average molecular weight of Polymer B was about 3000 g/mol, and of a second batch, Polymer B', was about 10000 g/mol. These results are in agreement with molecular weights measured by gel permeation chromatography, and no indication of cyclic oligomers were observed.

Differential scanning calorimetry showed that both of these polymers undergo two glass transitions, a low one at about 18°C to 20°C (likely due to phase separation of the fatty chains), and a higher one at about 92° for Polymer A and about 66° for Polymer B' (Figure 2).

Film Deposition

Chloroform solutions of Polymer A (typically 0.5 mg/ml) and chloroform/pyridine (90/5 v/v) solutions of Polymer B (or B') (typically 0.6 mg/ml) were spread on pure water (18 MegOhm, Millipore Milli-Q system) in the two-compartment NIMA film

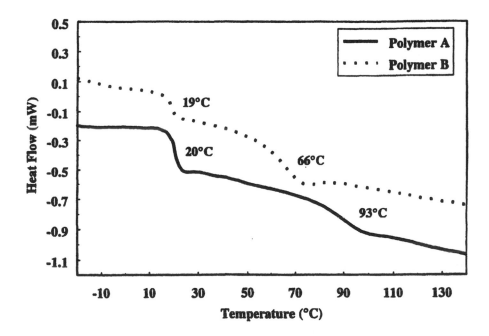

Figure 2 Differential scanning calorimetry: Polymer B' (top) and Polymer A (second heating scans @ 10°/minute under a nitrogen atmosphere).

balance. The solvents were allowed to dry from the films in each compartment for several minutes, then the films were compressed at a rate of about $20 \, cm^2/min$ until a surface pressure of about $20 \, mN/m$ was obtained. The compressed films were aged at $20 \, mN/m$ for about 20 min to allow densification of the monolayers and to assure that there was no leakage of polymer from the compression area. Multilayer films were deposited onto hydrophobic microscope slides at a rate of about $2-3 \, mm/min$. The glass slides were made hydrophobic by holding them in refluxing hexamethyldisilizane (HMDS) for about 30 min.

During the LBK deposition, the change in surface area of each compartment was monitored as the transfer of film was made at constant surface pressure. Polymer A, the first layer, was deposited on the down strokes, and Polymer B (or B') was deposited on the up strokes. Typically, the average deposition ratio of Polymer A (down stroke) was about 0.5–0.7, and the average deposition ratio of Polymer B (or B') (up stroke) was about 0.9–1.5. The reason for the large deviations from unity is unknown.

Optical Measurements

Linear absorption and second-harmonic generation (SHG) intensity measurements were made in transmission on freshly prepared films of various thicknesses. The linear absorption measurements were made normal to the films with a Cary 5E UV-VIS spectrometer. The SHG signal was generated by transmission of a fundamental beam from a Q-switched Nd:YAG laser (pulse width of 10 ns and repetition rate of 10 Hz) at an angle of about 30° from normal incidence to the film. These data are plotted in Figure 3. The SHG intensity was measured with an array of intensified Si-photodiodes (each point on Figure 3(b) is an average of 1500 laser pulses). The quadratic increase in SHG intensity as a function of thickness and the linear increase of absorption with thickness indicate the films have a high degree of uniformity of thickness and polar alignment.

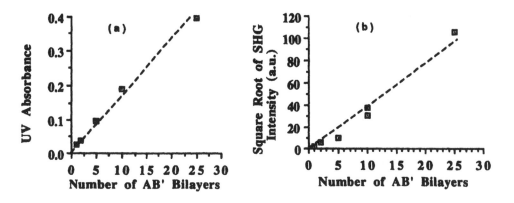

Figure 3 LBK bilayers of polymers A (1460-44) deposited on downstroke and B' (1500-22) deposited on upstroke. (a) Linear increase in the maximum UV absorbance with the number of bilayers; (b) Quadratic enhancement of SHG intensity with the number of bilayers.

The experimental setup for observing polarized SHG about the azimuthal angle for these films is as follows: The sample was at the waist of a focused (100 mm lens) beam (1064 nm) from a Q-switched Nd : YAG laser. The pulse width was 150 ns and the repetition rate was 1 kHz. The fundamental beam was polarized inside the laser cavity and a half-wave plate was used to control the polarization of the beam incident on the sample. The second-harmonic signal was detected by a Hamatsu R928 photomultiplier tube in conjunction with a Stanford Research SR250 boxcar averager using active baseline subtraction. To account for laser power fluctuations, the second harmonic was normalized by monitoring the fundamental signal. The sample was mounted on a Oriel rotation stage and data was collected at 5° increments of the azimuthal angle (ϕ). The angle of incidence of the laser beam was fixed at 56°.

Using a 24 bilayer film made from Polymer A and Polymer B, complete 360° SHG scans were recorded for 6 different polarization combinations of the fundamental and second harmonic. The s-polarized second harmonic from an s-polarized fundamental (s–s) was zero, and the p-polarized second harmonic from an s-polarized fundamental (p–s) was weak and nearly in the noise. The other 4 combinations (p–p, 45°–p, s–p, and 45°–s) gave significant second harmonic, and polar plots of these data are shown in Figure 4. Theoretical fits to these data included reflections of the second harmonic [10].

For this preliminary film, the long axis of the p–p oval is tilted 20° away from the dipping direction [11]. If the x axis is taken to lie along the long axis of the oval, the d-tensor which fits these data is:

$$
\begin{matrix}
0 & 0 & 0 & 0 & d_{31} & 0 \\
0 & 0 & 0 & d_{32} & 0 & 0 \\
d_{31} & d_{32} & d_{33} & 0 & 0 & 0
\end{matrix}
$$

Unlike the case for poled polymer films, d_{32} does not equal d_{31}.

For a chromophore with a single dominant hyperpolarizability component $\beta_{\xi\xi\xi}$ along the charge transfer axis x, the structure of d in the above d-tensor dictates [12] that the odd azimuthal moments are zero, leaving:

$$d_{33} \propto \langle \cos^3(\theta) \rangle \beta_{\xi\xi\xi},$$

$$d_{31} \propto \langle \cos(\theta)\sin^2(\theta) \rangle \langle \cos^2(\phi) \rangle \beta_{\xi\xi\xi},$$

$$d_{32} \propto \langle \cos(\theta)\sin^2(\theta) \rangle \langle \sin^2(\phi) \rangle \beta_{\xi\xi\xi},$$

where θ is the polar tilt angle of the chromophores' charge transfer axis ($\theta = 0°$ is normal to the film) and ϕ is the azimuthal angle of the chromophores' charge transfer axis ($\phi = 0°$ is parallel to the dipping direction). The results from Figure 3 has the chromophores' average polar tilt angle, θ, at about 45°. One model that fits the data in Figure 4 has the polymer chains in a helical conformation with the long axis of the helices aligned 20° on the average with respect to the dipping direction, and the chromophores tilted on both sides of the helical axis.

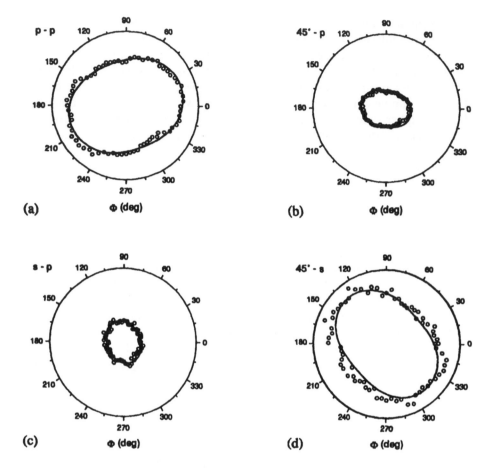

Figure 4 SHG Intensity (distance from center, a.u.) as a function of azimuthal angle: for (a) p–p and (b) 45°–p, the radial maximum = 0.4; for (c) s–p and (d) 45°–s, the radial maximum = 0.1. The fundamental laser beam was fixed at an angle of 56° from incidence.

ACKNOWLEDGMENTS

The authors are grateful to Dr. Rena Yee for thermal analysis, Dr. Larry Merwin for NMR analysis, Dr. Andrew Wright for linear absorption analysis, Dr. Eric Nickel for helpful discussion, and to the Office of Naval Research for financial support.

REFERENCES

1. P. N. Prasad and D. J. Williams, *Introduction to Nonlinear Optical Effects in Molecules and Polymers*, John Wiley & Sons, Inc: New York, 1991.
2. D. M. Burland, R. M. Miller and C. A. Walsh, *Chem. Rev.* **94**, 31–75 (1994).

3. G. L. Gaines, *Insoluble Monolayers at Liquid–Gas Interfaces*, Interscience: New York, 1966.
4. R. C. Hall, G. A. Lindsay, B. Anderson, S. T. Kowel, B. G. Higgins and P. Stroeve, *Mat. Res. Soc. Symp. Proc.* **109**, 351–356 (1988).
5. G. A. Lindsay, J. D. Stenger-Smith, R. A. Henry, J. M. Hoover, R. A. Nissan and K. J. Wynne, *Macromolecules* **25**(22), 6075 (1992).
6. J. M. Hoover, R. A. Henry, G. A. Lindsay, S. F. Nee, J. D. Stenger-Smith, in: *Organic Materials for Non-linear Optics III*, G. J. Ashwell and D. Bloor, Eds., The Royal Society of Chemistry: London, 1993; Special Publication 137.
7. T. L. Penner, *et al.*, *Nature* **367**, 49 (1994).
8. Synthesis of the electron donating bridge for polymer B:

Step 1 **N,N'-Bis-(2-hydroxyethyl)-N,N'-diphenylethylenediamine:** To a stirred solution of 1.1 g of LiAlH$_4$ (28 mmole) in 100 ml of Et$_2$O was slowly added (20 min) 3.15 g (8.2 mmole) of N,N'-*bis*(carboethoxymethyl)-N,N'-diphenylethylene-diamine in 100 ml of Et$_2$O. After stirring at room temperature for 3 h there was slowly added 1 ml H$_2$O followed by 1 ml 15% aq. KOH and finally 3 ml H$_2$O. After stirring for 30 min ca. 5 g of MgSO$_4$ was added with stirring and the mixture was filtered and the white solid washed with Et$_2$O (2 × 50 ml). The combined filtrates were evaporated at reduced pressure and the resulting solid recrystallized from ca. 40 ml MeOH giving 296 mg of N,N'-diphenylpiperazine, a contaminant in the starting material. The filtrate was evaporated at reduced pressure and the resulting solid recrystallized from benzene/cyclohexane giving tan cubic crystals (2.13 g; 85% yield). ^{1}H NMR (200 MHz; DMSO-d$_6$): 7.14 (4H, m), 6.69 (4H, m), 6.52 (2H, m), 4.69 (2H, br s), 3.48 (8H, br s), 3.35 (4H, t, J = 6.3 Hz).

Step 2 **N,N'-Bis-(2-acetoxyethyl)-N,N'-diphenylethylenediamine:** A mixture of 2 g (6.67 mmole) of N,N'-bis-(2-hydroxyethyl)-N,N'-diphenylethylenediamine and 300 mg of pyridine in 5 ml of Ac$_2$O was heated at ca. 95°C for 1.5 h and then evaporated at reduced pressure. The residue (an oil) was dissolved in 100 ml of CH$_2$Cl$_2$ and extracted with water, dried (MgSO$_4$) and evaporated at reduced pressure giving 2.72 g of a yellow oil. The product was quickly chromatographed (silica gel/CHCl$_3$) and after evaporation of solvent yielded a pale yellow oil (2.48 g, 97% yield), which solidified to a white solid. M.p. 59.5–60.5°C. ^{1}H NMR (200 MHz, acetone-d$_6$): 7.19 (4H, m), 6.83 (4H, m), 6.64 (2H, m), 4.19 (4H, t, J = 6 Hz), 3.64 (4H, t, J = 6 Hz), 3.63 (4H, s), 1.96 (6H, s).

Step 3 **N,N'-Bis-(2-acetoxyethyl)-N,N'-di(4-formylphenyl)-ethylenediamine:** A mixture of 5.9 g of dry DMF and 9.9 g of POCl$_3$ was heated to 70° and stirred for 10 min. To this mixture was added 2.42 g (6.3 mmole) of N,N'-bis-(2-acetoxyethyl)-N,N'-diphenylethylenediamine in 6 ml of 1,2-dichloroethane with stirring. The mixture was then heated to ca. 85°C and stirred for 2 h. After cooling, the reaction mixture was poured into a mixture of 250 ml CHCl$_3$ and 250 ml water and then stirred for 2 h. The layers were separated and the aqueous layer extracted with CHCl$_3$ (2 × 200 ml), the organic fractions were combined, washed with water (3 × 100 ml), dried (MgSO$_4$) and evaporated at reduced pressure. The residue was chromatographed (silica gel/CHCl$_3$) giving a yellow oil, which slowly solidified. Recrystallization from MeOH/H$_2$O gave tan cubic crystals (1.74 g, 63% yield). M.p. 121.5–2°C. ^{1}H NMR (200 MHz, acetone-d$_6$): 9.74 (2H, s), 7.73 (4H, m), 6.96 (4H, m), 4.26 (4H, t, J = 6 Hz), 3.87 (4H, s), 3.79 (4H, t, J = 6 Hz), 1.96 (6H, s).

Step 4 **N,N'-Bis-(2-hydroxyethyl)-N,N'-di(4-formylphenyl)ethylenediamine:** A mixture of 440 mg (1 mmole) of N,N'-bis-(2-acetoxyethyl)-N,N'-di(4-formyl-phenyl)-ethylenediamine, 40 drops of conc. H_2SO_4 and 10 ml H_2O in 100 ml CH_3OH was heated at reflux for 4 h. After cooling, a few pieces of ice and a few ml of water were added and the mixture was then cooled in an ice bath. The pale yellow solid that separated was filtered, washed with water and dried giving a pale yellow fine crystalline solid (282 mg, 64% yield). M.p. 204–5°C. Repeated concentration of the filtrate and recooling yielded a second (46 mg) and third crop (39 mg) of product. A small sample of material was recrystallized from benzene/MeOH giving pale yellow crystals. M.p. 211–212°C. ^1H NMR (200 Mhz, DMSO-d_6): 9.66 (2H, s), 7.66 (4H, m), 6.84 (4H, m), 3.70 (4H, b s), 3.51 (10H, m).

Synthesis of the electron accepting bridge for polymer B:

Step 1 **Ethylenediamine dilauramide:** 28 ml Lauryl chloride (120 mmoles) was added dropwise to a solution of 4 ml ethylenediamine (60 mmoles) in 250 ml pyridine over 15 min. The mixture was heated to 100°C then allowed to cool overnight. 400 ml Hexanes were added and the solids filtered off and washed with hexanes. After drying, the solids were stirred for 10 min with 200 ml 4N HCl. The solids were then filtered off and dried *in vacuo* to give 17.86 g (70%) of a white powder. This can be recrystallized from 400 ml of 1 : 1 ethanol/toluene to give 17.16 g of a white powder. ^1H NMR (DMSO, 90°C) 3.12 (4H, t), 2.05 (4H, t), 1.50 (4H, t), 1.26 (36H, m), 0.87 (6H, t). IR (cm^{-1}) 2910 (s), 1655 (s), 1570 (s).

Step 2 **N,N'-*bis*dodecylethylenediamine:** 17.16 g solid ethylenediamine dilauramide (40 mmoles) was added in portions to a stirred suspension of 4.60 g LiAlH$_4$ (121 mmoles) in 250 ml anhydrous THF under N_2 with cooling (water bath). The mix was refluxed for 48 h then cooled and quenched by the careful addition of 35 ml 4M NaOH. The white solid was filtered off and washed with THF then discarded. The filtrate was concentrated in vaccum to yield 15.72 g (99%) of a waxy white solid. 11.75 g of this material was recrystallized from 20 ml H_2O/150 ml ethanol to give 10.38 g of a white waxy solid (88% recovery). ^1H NMR (CDCl$_3$): 2.78 (4H, s), 2.63 (4H, t), 1.55 (4H, t), 1.28 (40H, m) 0.88 (6H, t). ^{13}C NMR (CDCl$_3$): 49.78, 48.83, 31.90, 29.61–29 (overlapping peaks), 27.30, 22.67, 14.10.

Step 3 **N,N'-*bis*dodecylethylenediamine *bis*-chloroacetamide:** 0.4 ml chloroacetyl chloride (5 mmoles) was added dropwise to a solution of 1.98 g N,N'-*bis*-dodecylethylenediamine (5 mmoles) in 100 ml chloroform. After 30 min 0.7 ml triethylamine (5 mmoles) was added followed by another 0.4 ml chloroacetyl chloride. This procedure was repeated after another 30 min. The mixture was stirred overnight then washed with water, 4N HCl and finally by sat. aq. NaHCO$_3$. The solution was dried with MgSO$_4$ and concentrated in vacuum to yield 2.52 g (92%) of a thick straw colored oil. This was used as is in the following procedure. ^1H NMR (CDCl$_3$): 4.04 (4H, s), 3.80 (4H, s), 3.29 (4H, t), 1.6 (4H, t), 1.23 (40H, m), 0.85 (6H, t).

Step 4 **N,N'-*bis*dodecylethylenediamine *bis*-cyanoacetamide:** A mixture of 2.26 g (4.1 mmoles) N,N'-*bis*dodecylethylenediamine *bis*-chloroacetamide and 0.5 g (10 mmoles) powdered NaCN in 50 ml DMSO was heated to 100°C and held there for three days. The solution was cooled and poured into 600 ml cold water. The aq. solution was extracted with 300–100 ml ether. The combined extracts were washed well with water then dried over MgSO$_4$, filtered and concentrated in vacuum to give 1.85 g (85%) of a brown glassy solid. This was chromatographed on Silica Gel eluting with 40% ethyl acetate/hexanes to yield 0.78 g (36%) of a brown glassy material. ^{1}H NMR (CDCl$_3$): 3.51 (4H, s), 3.49 (4H, s), 3.24 (4H, t), 1.29 (40H, m), 0.88 (6H, t).

9. The assignments for the 400 MHz ^{1}H NMR spectrum of polymer A in CDCl$_3$ solution are as follows (see letters on Figure 1 corresponding to each resonance): (a) resonance at δ 8.1 (2H vinylic); (b) resonance at δ 7.9 (4H aromatic ortho to vinyl); (c) resonance at δ 7.1 (2H amide NH); (d) resonance at δ 6.7 (4H aromatic ortho to amine); (e) resonance at δ 4.0 (1H methine); (f) resonance at δ 3.6 (8H methylene alpha to amine); (g) resonance at δ 3.5 (1H hydroxyl); (h) resonance at δ 3.3 (4H methylene alpha to amide); (i) resonance at δ 1.6 (4H methylene beta to amine); (j) resonance at δ 1.3 (52H aliphatic methylene); and resonance at δ 0.90–0.85 (6H methyl). [The preliminary NMR assignments for Polymer A were given in Ref. 5 (as polymer XIV).] The assignments for the 200 MHz ^{1}H NMR spectrum of polymer B (and B') are as follows (see letters on Figure 1 corresponding to each resonance): (a) resonance at δ 7.8–8.1 (6H vinylic and aromatic meta to the amine); (b) resonance at δ 6.7–7.05 (4H aromatic ortho to amine); (c) resonance at δ 6.4–6.6 (2H hydroxyl); (d) resonance at δ 3.8–4.2 (12H protons on carbon alpha to amine nitrogen and protons on carbons alpha to alcohol oxygen); (e) resonance at δ 3.4–3.8 (8H protons on carbon alpha to amide nitrogen); (f) resonance at δ 1.6–1.9 (4H protons on carbon beta to amide nitrogen); (g) resonance at δ 1.0–1.5 (32 H methylene aliphatic group); (h) resonance at δ 0.7–0.90 (6H methyl). Aldehyde end groups were detectable at 9.75 ppm.

10. W. N. Herman and L. M. Hayden, *J. Opt. Soc. Am. B* **12**, 416 (1995).

11. Subsequent films, made under slightly different conditions, showed no preferential in-plane order.

12. M. B. Feller, W. Chen, and Y. R. Shen, *Phys. Rev. A* **43**, 6778 (1991).

4a. TWO-DIMENSIONAL CHARGE-TRANSFER MOLECULES FOR POLED POLYMER APPLICATIONS

TATSUO WADA[a], TAKASHI ISOSHIMA[a], TETSUYA AOYAMA[b], YADONG ZHANG[a], LIMING WANG[a], JEAN-LUC BRÉDAS[a,c], KEISUKE SASAKI[b], and HIROYUKI SASABE[a]

[a]Frontier Research Program, The Institute of Physical and Chemical Research (RIKEN), 2-1 Hirosawa, Wako, Saitama 351-01, Japan; [b]Department of Material Sciences, Faculty of Science and Technology, Keio University, 3-14-1 Hiyoshi, Kohoku-ku, Yokohama 223, Japan; [c]Service de Chimie des Matériaux Nouveaux et Centre de Recherche en Electronique et Photonique Moléculaires, Université de Mons-Hainaut Place du Parc, 20, B-7000 Mons, Belgium

ABSTRACT

The advantage of two-dimensional (2D) charge-transfer (CT) molecules in a poled polymeric system over one-dimensional (1D) CT ones is experimentally and theoretically demonstrated for waveguide device applications. Significant enhancements of off-diagonal second-harmonic and electro-optic tensor components were observed in poled polymers doped with 2D CT molecules. Poling behavior of second harmonic generation coefficient and linear polarizability tensor components were calculated for carbazole derivatives with both 2D and 1D CT character. It is shown that the off-diagonal components for a 2D CT molecule are 1.5 times as large value as those in 1D CT molecules. In addition, the 2D CT molecule has less anisotropy refractive index than the 1D CT molecule, which is preferable for waveguide applications.

INTRODUCTION

Organic charge-transfer (CT) molecules have been experimentally and theoretically shown to have anomalously large second-order nonlinear optical susceptibilities [1]. Their crystals and polymeric systems have been developed to exhibit efficient second-order nonlinear optical responses [2]. Among them, poled polymers are one of the promising systems because of thin film processabilities and compatibility for waveguide fabrication. The temporal and thermal stability of electro-optic responses in poled polymers have been demonstrated by various approaches. Thermally stable guest/host systems have large flexibility for material design and have been applied to waveguide devices. Linear and nonlinear optical properties of poled polymers are determined by the nature of the constituent second-order nonlinear optical chromophores (NLO-phores) as well as the geometrical relationship between the dominant CT transition dipole vector and the ground-state dipole moment vector. The former is the origin of molecular hyperpolarizability and the latter can be aligned by a poling electric field. Although organic chemistry allows us to synthesize various kinds of CT molecules, one-dimensional (1D) CT molecules are the well-studied and typical NLO-phores for poled polymers. In 1D molecules, donor and acceptor groups are incorporated at the opposite ends of a π-conjugation path to enhance the diagonal

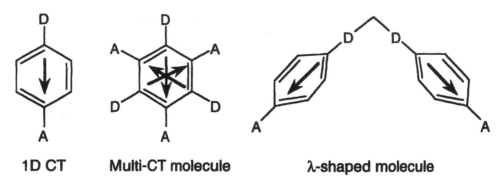

1D CT   **Multi-CT molecule**   **λ-shaped molecule**

Figure 1   Various types of charge-transfer (CT) molecules.

tensor component as shown in Figure 1. Recently, Miyata *et al.* [3] developed the
λ-shaped molecules for phase matched second harmonic generation (SHG) in
crystals and Zyss *et al.* [4] revealed the advantages for electro-optic applications of
multi-CT molecules such as octupolar molecules over 1D molecules in Figure 1. In
order to achieve high and stable performance in waveguide device applications, the
NLO-phores should be designed to fulfill various requirements besides large nonlinearity and small orientational relaxation. For example utilization of off-diagonal
components in SHG tensor provides more flexibility in phase matching method;
refractive index change due to relaxation should be suppressed since most of
waveguide devices are very sensitive to the change in refractive index of materials.
Taking the above discussion into account our efforts have been focussed on the
development of novel multifunctional NLO-phores suitable for thermally stable
poled polymer applications. In this paper we describe linear and nonlinear optical
behavior in two-dimensional (2D) CT molecules experimentally and theoretically.

2D CT CHARACTER IN DI-SUBSTITUTED CARBAZOLE

In searches for multifunctional NLO-phores we have developed various carbazole
derivatives which possess both photoconductivity and second-order nonlinear optical
response [5]. Carbazole compounds are well known to exhibit good hole transporting
properties and their photocarrier generation efficiency can be sensitized by formation
of CT complexes [6]. The carbazole molecule has an isoelectronic structure of
diphenylamine, that is, electron-donative nature as shown in Figure 2. Therefore, the
introduction of electron-withdrawing groups in the 3 and/or 6-position induces intra-
charge-transfer. Depending on the electron-affinity of acceptor groups, the additional
CT bands are superimposed in a visible region. Various substituent groups were syste-
matically examined in different positions such as R_1, R_2 and/or R_3 in Figure 2. Although
9-substituent group does not significantly affect the electronic property of carbazole
rings, the hydrogen bonding in 9-hydroxyethyl substituent plays the important role of
generating noncentrosymmetric packing of carbazole molecules in a crystal [7].

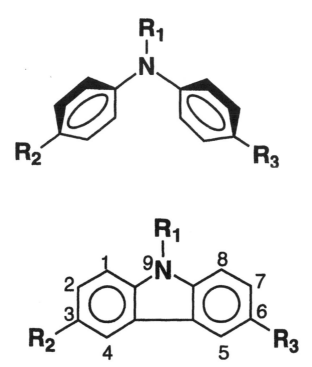

Figure 2 Substituted diphenylamine and substituted carbazole structures.

The single crystals of 9-hydroxyethylcarbazole and 3-nitro-9-hydroxyethylcarbazole were determined to have noncentrosymmetric crystallographic structures with tetragonal ($I4_1$) and monoclinic ($P2_1$) crystal systems. Therefore one can expect that not only optical properties but also packing properties can be controlled by proper molecular design of substituent groups in substituted carbazole derivatives.

Di-substituted carbazoles such as 3,6-dinitrocarbazole derivatives have similar electronic properties as di(4-nitrophenyl) amine. Unlike 3-nitro-9-hydroxyethylcarbazole crystal, 3,6-dinitro-9-hydroxyethylcarbazole has a centrosymmetric unit cell with space group $P\bar{1}$. Although dinitrocarbazole crystal shows no SHG activity, one of the interesting features is that this di-substituted carbazole has two CT axes. Therefore, the vector direction of the net ground-state dipole moment is not parallel to that of molecular CT axis. As a consequence, CT contribution to the nonlinear hyperpolarizabilities of this molecule has an intrinsic 2D character. Therefore, this dinitrocarbazole molecule can be expected to show different behavior under the electric-field induced alignment of the net ground-state dipole moment.

Figure 3 indicates the results of theoretical calculations [8]: the optimized molecular geometry was obtained by the AM1 calculation using MOPAC and the INDO-SCI calculations were made to estimate the excited states. The charge-transfer

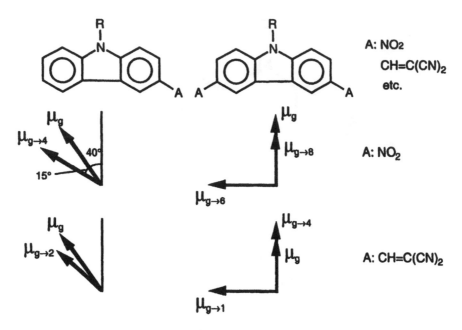

Figure 3 Molecular structures of mono- and di-substituted carbazoles. The bottom shows the directions of the ground-state and transition dipole moments.

in dicyanovinylcarbazole derivative has a quasi-one-dimensional character in which the ground-state dipole moment (μ_g) and the charge-transfer transition dipole moment ($\mu_g \to 4$ or $\mu_g \to 2$) are almost parallel. On the other hand, di-substituted carbazole has two CT axes which induce two dominant CT transitions. The theoretical results indicate that there is a perpendicular component ($\mu_g \to 6$ or $\mu_g \to 1$) of CT transition dipole moments in addition to the parallel component to μ_g.

ELECTRIC-FIELD INDUCED ALIGNMENT OF 2D CT MOLECULES

Electric-field induced alignment (poling) is widely used to break the centrosymmetry in a guest–host, side-chain polymer, or main-chain polymer system. The ground-state dipole moment in NLO-phores can be aligned by an electric field. Linear and non-linear optical properties after poling are determined by the relative geometrical relationship between CT transition dipole moments and ground-state dipole moment. Therefore, it is necessary to elucidate this relationship in 2D CT molecules which is not well studied.

Figures 4 and 5 show the measured (thick lines) and calculated (thin lines) absorption spectra along with a schematic picture of the electric-field induced alignment of molecules [9]. Full and dotted lines correspond to the spectra before and after poling, respectively. Although there are some differences in wavelength, the key features of the measured spectra such as relative peak positions and intensities are rather well

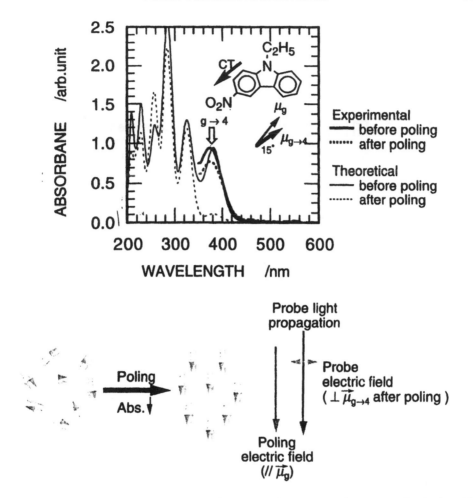

Figure 4 Measured (thick lines) and calculated (thin lines) absorption spectra for 3-nitro-N-ethylcarbazole. Full and dotted lines are the absorption spectra before and after poling, respectively.

reproduced by the calculation results. The insets show the arrangement of the ground-state dipole moment (μ_g) and the CT transition dipole moments ($\mu_g \rightarrow 4$ or $\mu_g \rightarrow 6$) for those absorption peaks. The poling behavior in absorption changes also agrees with the theory. The decrease of the longest-wavelength absorption peak was obviously observed in mono-substituted derivatives independent of acceptor groups. On the other hand, di-substituted derivatives showed a complicated behavior of absorption changes after poling depending on the acceptor groups. In the di-substituted derivatives, there are two large CT transitions as can be seen in Figure 5. The arrangement of the ground-state and transition dipole moments are quite different from the mono-substituted derivatives, as is shown in Figure 3. The ground-state dipole moment can be expressed, in a very rough approximation, as

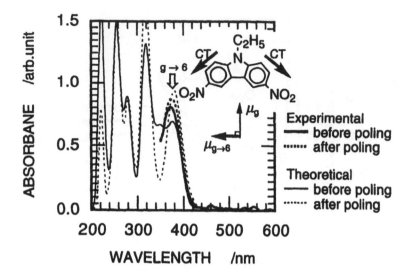

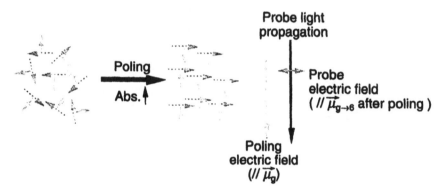

Figure 5 Measured (thick lines) and calculated (thin lines) absorption spectra for a 3,6-dinitrocarbazole derivative. Full and dotted lines are the absorption spectra before and after poling, respectively.

vector sum of dipole moments for each substitution, and therefore almost parallel to the symmetry axis of the molecule in the case of symmetrically di-substituted carbazole. The two CT systems are closely coupled, and two CT transitions are nondegenerate, as is shown in Figure 5. The lowest transition has a transition dipole moment vector perpendicular to the symmetry axis, and the next lowest transition has a transition dipole moment vector parallel to the symmetry axis, which is consistent with the consideration on the symmetry of the coupled transitions. Both contribute to the absorption before poling, and only the lowest contributes after perfect poling. Whether the peak increases or decreases depends on the angle between the CT axes, which can vary with the acceptor groups besides and the conformational isomers.

ENHANCEMENT OF OFF-DIAGONAL TENSOR COMPONENTS IN SHG AND ELECTRO-OPTIC EFFECTS

We demonstrated the enhancement of off-diagonal tensor components in poled polymers doped with 2D CT molecules by SHG and electro-optic measurements. In poled polymers containing 1D CT molecules such as p-nitroaniline and p-substituted stilbene, the diagonal SHG tensor component d_{33} is three times larger than the other component d_{31} under a low electric field [10]. These components, d_{33} and d_{31}, can be determined independently by measuring p-polarized second-harmonic intensities from the poled polymer films with s- and p-polarized fundamental light, respectively. Poly(methyl methacrylate) (PMMA) doped with 2D CT molecules such as 3,6-dinitro-9-heptylcarbazole was determined to possess a $d_{33} = d_{31}$ relationship after corona-poling as shown in Figure 6 [5]. Thus 2D CT molecules can be utilized as a unique NLO-phore to improve the nonlinear optical properties of poled polymers, especially an enhancement of the off-diagonal tensor component d_{31}, which provides the flexibility of a device design for phase matched SHG.

Besides the reflection method [11,12] the Mach–Zehnder method [13] was performed to determine the electro-optic coefficients r_{33} and r_{13} independently for dicyanovinyl and di(dicyanovinyl)carbazoles. The electro-optic coefficients measured at a wavelength of 632.8 nm are summarized in Table 1 [14]. We repeated the measurement on the samples which have different degrees of alignment by poling. It should be noticed that the tensor ratios are almost same in the same samples with different degrees of poling. The ratios of the electro-optic coefficients are around 3

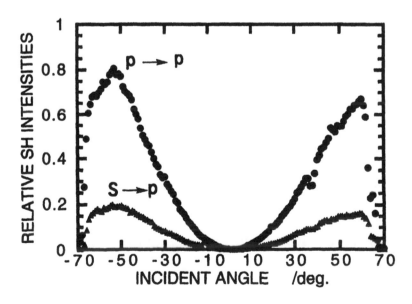

Figure 6 Relative second-harmonic intensities as a function of incident angle in PMMA doped with a 3,6-dinitrocarbazole derivative.

Table 1 Electro-optic coefficients measured by the Mach–Zehnder method and the reflection method (r_{33}^{ref}) for (a) PMMA doped with 9.1 wt% of N-ethyl-3-dicyanovinylcarbazole; and (b) PMMA doped with 2.0 wt% of N-heptyl-3,6-di(dicyanovinyl) carbazole.

r_{33}^{ref} (pm/V)	r_{13} (pm/V)	r_{33} (pm/V)	r_{33}/r_{13}
a 0.68	0.25	0.77	3.1
0.63	0.28	0.81	2.9
0.21	0.044	0.13	3.0
b 0.25	0.18	0.29	1.6
0.19	0.23	0.32	1.4
0.12	0.12	0.16	1.4

for corona-poled PMMA films doped with 9.1 wt% of dicyanovinyl carbazole which has quasi-1D CT character. On the other hand, those for corona-poled PMMA films doped with 2.0 wt% of di(dicyanovinyl) carbazole are around 1.5. These differences in the tensor ratios observed in SHG and electro-optic effects are presumably due to the difference between the 1D CT character of mono-substituted carbazole and 2D CT character of di-substituted carbazole.

LINEAR AND NONLINEAR OPTICAL ANISOTROPY IN 2D CT MOLECULES

The tensor components of the molecular hyperpolarizability were obtained by the sum-over-states method. The tensors were transformed into a Cartesian coordinates system in which the molecular ground-state dipole moment vector μ_g is z-axis (μ_g-coordinates), and were averaged over rotation around z-axis. The macroscopic molecular hyperpolarizability tensors were obtained by transforming into a Cartesian coordinates system in which the poling electric field E_{pol} is along the z-axis (macroscopic coordinates), and then by averaging over the direction of μ_g weighted by the Boltzmann factor. The hyperpolarizability tensor components are expressed as follows [5]:

$$\langle \beta_{31} \rangle = \beta_{31} \frac{L_1(p) + L_3(p)}{2} + (\beta_{33} - 2\beta_{15}) \frac{L_1(p) - L_3(p)}{2},$$

$$\langle \beta_{33} \rangle = \beta_{33} L_3(p) + (\beta_{31} + 2\beta_{15})(L_1(p) - L_3(p)),$$

$$\langle \beta_{15} \rangle = \beta_{15} L_3(p) + (-\beta_{31} + \beta_{33}) \frac{L_1(p) - L_3(p)}{2},$$

where $L_i(p)$ is the ith order Langevin function, and p is the poling factor $|\mu_g\|E_{pol}|/kT$. Under the condition of small p, a well-known relation, $\langle\beta_{33}\rangle = 3\langle\beta_{31}\rangle$, is derived assuming Kleinman's symmetry $\beta_{15} = \beta_{31}$. In the case of large p, $\langle\beta_{31}\rangle$, $\langle\beta_{33}\rangle$, and $\langle\beta_{15}\rangle$ asymptotically approach β_{31}, β_{33}, and β_{15}, respectively, as expected. The linear polarizability tensor components are expressed as follows:

$$\langle\alpha_{11}\rangle = \alpha_{11}\frac{1 + L_2(p)}{2} + \alpha_{33}\frac{1 - L_2(p)}{2},$$

$$\langle\alpha_{33}\rangle = \alpha_{11}(1 - L_2(p)) + \alpha_{33}L_2(p).$$

The macroscopic SHG coefficient d_{ij} and refractive index n_i are obtained from these macroscopic molecular hyperpolarizability tensors by multiplying the number density of molecules and the local field factors using an oriented gas model description. The molecular hyperpolarizability tensor components in the μ_g-coordinates at 1064 nm are $\alpha_{11} = 14.1$, $\alpha_{33} = 31.2$ [$\times 10^{-24}$ esu], $\beta_{31} = -3.5$, $\beta_{33} = 99.8$, and $\beta_{15} = -1.0$ [$\times 10^{-30}$ esu] for nitrocarbazole, and $\alpha_{11} = 21.1$, $\alpha_{33} = 25.5$ [$\times 10^{-24}$ esu], $\beta_{31} = 28.7$, $\beta_{33} = 31.3$, and $\beta_{15} = 34.4$ [$\times 10^{-30}$ esu] for dinitrocarbazole. In the 2D CT molecule, the off-diagonal components are significantly enhanced at the cost of a decrease of the diagonal component. In addition, the molecular polarizability is less anisotropic in the 2D CT molecule than in the 1D CT molecule [15]. These enhanced off-diagonal tensor components were obtained in λ-shaped molecules [16].

Figure 7 shows the evolution of the SHG tensor components with the poling factor p. In the corona-poling method, the accessible poling electric field is on the order of 10^6 V/cm. Thus, p is as large as 3 for the molecules with ground-state dipole moment of several Debye. At this value of p, the off-diagonal components in the 2D CT molecule are about 1.5 times as large as those in the 1D CT molecule although the 2D and 1D CT molecules behave in a similar manner at small p value. Watanabe et al. [16] reported that the changes in nonlinear optical coefficients for λ-shaped molecules were suppressed under the relaxation of order parameter of dipole because the decrease of diagonal tensor components were compensated with the off-diagonal tensor components.

Figure 8 shows the evolution of the refractive index of PMMA doped with 10 wt% of carbazole derivatives. These theoretical values present good agreement with the experimental value for an unpoled nitrocarbazole/PMMA sample, 1.520 at 632.8 nm, obtained by the mode-line measurement. In the 2D CT system, birefringence induced by poling is much smaller than in the 1D CT system, because of less anisotropy in the molecular linear polarizability. This is a great advantage of 2D CT systems over 1D CT systems for waveguide device applications, since these devices are very sensitive to the refractive index of the material. For example, in an electro-optically modulated directional coupler, coupling length is very sensitive to the refractive index. Similarly in an SHG device, the phase matching condition is strongly affected by the refractive index through the coherence length. Figure 9 shows the evolution of coherence length for SHG at 1064 nm in PMMA doped with 10 wt% of carbazole derivatives. It should be noted that the scale for vertical axis is different between the

T. WADA *et al.*

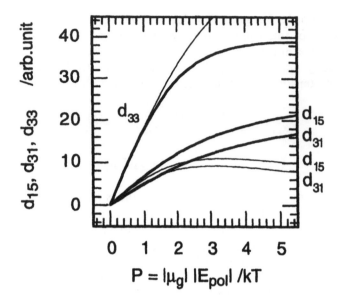

Figure 7 Evolution of SHG tensor components with poling factor *p*. Thick lines represent the case of 2D CT 3,6-dinitrocarbazole and thin lines represent the case of 1D CT 3-nitrocarbazole.

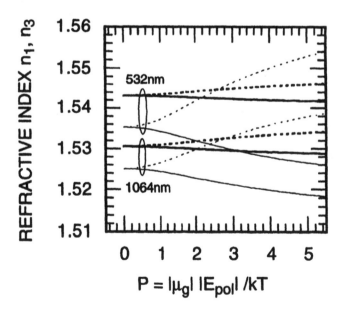

Figure 8 Evolution of refractive indices n_1 and n_3 with poling factor *p*. Full lines represent the refractive indices for TE polarization n_1, and dashed lines represent that for TM polarization n_3. Thick lines represent refractive indices of 2D CT 3,6-dinitrocarbazole, and thin lines represent those of 1D CT 3-nitrocarbazole.

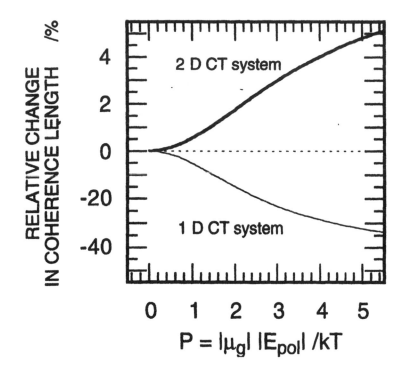

Figure 9 Evolution of coherence length for SHG in PMMA doped with 10 wt% of carbazole derivatives with poling factor p. Fundamental wavelength is 1064 nm and TE polarization is assumed in fundamental and SH field. Coherent length at $p = 0$ is 21 μm for 2D CT 3,6-dinitrocarbazole and 26 μm for 1D CT 3-nitrocarbazole.

2D and 1D CT systems. The relative change in the 2D CT system is nearly one order of magnitude smaller than the change in the 1D CT system. Hence the 2D CT molecule significantly improves the tolerance for orientational relaxation in waveguide devices.

REFERENCES

[1] S. J. Lalama and A. F. Garito, *Phys. Rev. A*, **20** (1979) 1179.
[2] L. R. Dalton, A. W. Harper, R. Ghosn, W. H. Steier, M. Ziari, H. Fetterman, Y. Shi, R. V. Mustacich, A. K.-Y. Jen and K. J. Shea, *Chem. Mater.*, **7** (1995) 1060.
[3] H. Yamamoto, S. Katogi, T. Watanabe, H. Sato and S. Miyata, *Appl. Phys. Lett.*, **60** (1992) 935.
[4] I. Ledoux, J. Zyss, J. S. Siegel, J. Brienne and J.-M. Lehn, *Chem. Phys. Lett.*, **172** (1990) 440.
[5] T. Wada, Y. Zhang, M. Yamakado and H. Sasabe, *Mol. Cryst. Liq. Cryst.*, **227** (1993) 85.
[6] H. Hoegl, *J. Phys. Chem.*, **69** (1965) 755.
[7] T. Wada, Y. Zhang, Y. S. Choi and H. Sasabe, *J. Phys. D: Appl. Phys.*, **26** (1993) B221.

[8]   T. Isoshima, Y.-D. Zhang, E. Brouyère, J.-L. Brédas, T. Wada and H. Sasabe, *Nonlinear Opt.*, **14** (1995) 175.

[9]   T. Isoshima, T. Wada, Y.-D. Zhang, E. Brouyère, J.-L. Brédas and H. Sasabe, *J. Chem. Phys.*, **104** (1996) 2467.

[10] K. D. Singer, M. G. Kuzyk and J. E. Sohn, *Nonlinear Optical and Electroactive Polymers*, Eds. P. Prasad and D. Ulrich (Plenum Press, New York, 1988), pp. 189–204.

[11] C. C. Teng and H. T. Man, *Appl. Phys. Lett.*, **56** (1990) 1734.

[12] J. S. Schildkraut, *Appl. Opt.*, **29** (1990) 2839.

[13] R. A. Norwood, M. G. Kuzyk and R. A. Keosian, *J. Appl. Phys.*, **75** (1994) 1869.

[14] T. Aoyama, T. Wada, Y.-D. Zhang, T. Isoshima, Y. Togo, H. Sasabe and K. Sasaki, *Nonlinear Opt.*, **15** (1996) 403.

[15] T. Isoshima, T. Wada, Y.-D. Zhang, T. Aoyama, J.-L. Brédas and H. Sasabe, *Nonlinear Opt.*, **15** (1996) 65.

[16] T. Watanabe, M. Kagami, H. Miyamoto, A. Kidoguchi and S. Miyata, *Proc. of the Fifth Toyota Conference on Nonlinear Optical Materials* (Elsevier Science Publishers, 1992), pp. 201–206.

5. STABLE SECOND-ORDER NONLINEAR OPTICAL PROPERTIES OF ORGANIC COMPOUNDS IN SILICA MATRICES

JAE-SUK LEE[a], GYOUJIN CHO[b], and YONG GUN SHUL[c]

[a]Dept. of Materials Sci. and Eng., Kwangju Institute of Sci. and Tech. (K-JIST), Kwangju, Korea; [b]Dept. of Chem. and Biochem., Univ. of Oklahoma, Oklahoma, USA; [c]Dept. of Chem. Eng., Yonsei Univ., Seoul, Korea

ABSTRACT

A sol–gel process can offer stable silica matrices for organic chromophores possessing nonlinear optical (NLO) properties. In particular, DR1, carbazole, and MNA chromophores have been physically incorporated into the silica matrices. In addition, modified DR1 (DRS) have also been chemically incorporated into such silica matrices. The thermal stability of both physically and chemically incorporated systems has been investigated by UV/VIS spectroscopy. It reveals an enhanced thermal stability with the chemically incorporated system (DRS/silica) compared to that of the physically incorporated system. In addition, second harmonic generation (SHG) responses have additionally been examined for the DRS incorporated system.

INTRODUCTION

The origin of nonlinear optics may be directly traced to the invention of the laser. Both the high intensity and the degree of coherence of a nonlinear optical apparatus facilitate the detectable occurrence of multiphoton second, third, and even higher-order optical effects in matter. In recent years, as photonics materials have gained much attention as they have been increasingly required for the development of technology in information and communications processing [1], materials capable of generating new electromagnetic fields and subsequently dealing with their resultant interactions when applied. The field of nonlinear optics is therefore expected to become both a novel and key technology in related fields. In general, the rapid progress presently being achieved in these areas would be greatly enhanced by the preparation of readily processable, organic materials yielding sufficiently large responses [2]. Although more efficient photon manipulating materials in particular the utilization of molecularly based materials with complex microstructural arrangements are already a reality, the nonlinear optical materials commercially available are still limited in composition to inorganic crystals [3,4].

Recent published results, however, imply that molecular, organically based, macroscopic-electron assemblages possess several desirable NLO responses. Specifically, these materials offer ultra-fast response times, lower dielectric constants, better processing abilities, and enhanced NLO responses when compared to the traditional inorganic solids [2].

To be considered for most practical applications, electro-optic organic materials must possess both noncentrosymmetry in their macroscopic assemblies as well as a sufficient compatability with the processing methods and inherent operating environments encountered at high temperatures. Detailed investigations have therefore been recently attempted to yield the more accessible organic NLO materials: such investigations have typically consisted of either physically or chemically incorporating certain organic guest chromophores into host polymers [5], cross-linked polymers [6], and sol–gel thin films [7] to subsequently obtain the desired noncentrosymmetry by electric poling techniques.

Preliminary results reported in the literature on the physical incorporation of guest–host polymer systems have revealed that practical NLO property values may be obtained after applying poling techniques (Scheme 1(a)) [5]. The resultant incorporated chromophores then display distinct advantages over typical, physically incorporated guest–host polymer systems. Moreover, high NLO chromophores can actually be employed into a polymer system without the additional burdens of crystallization, phase separation, or even the formation of concentration gradients.

The host–guest system has been extended further by chemically attaching organic chromophores to the polymer backbone utilized (Scheme 1(b)). In addition, the relaxation of the poled order obtained was substantially slower obviously due to an increased hindrance of the chromophore in the polymer backbone. A typical example of such physical incorporation might therefore include a random copolymer comprised of poly(methyl methacrylate) based side chain polymer with a stilbene derivative and methacrylate, this system having already been widely investigated [8]. A poled-order relaxation of the type described above should logically be further

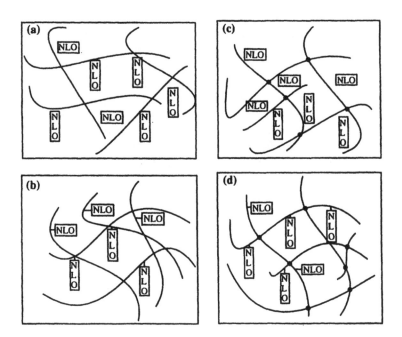

Scheme 1   Guest–host system using various organic NLO chromophores in polymer matrices systems (✛ : crosslinked point).

hindered if the organic chromophore utilized were to be chemically incorporated into the polymer backbone itself. The primary cause for such a lower, poled-order relaxation is commonly believed to be due to the large, segmental motions of the polymer backbone required for both poling and relaxation when the organic chromophore is used as the polymer backbone itself. A typical main chain polymer with these particular properties might be aromatic polyureas, which not only yields relatively high NLO coefficients displaying a color range from mostly transparent to light blue, but is also highly processable.

The poled-order relaxation processes which limit the lifetime of poled-polymer devices may be suppressed by incorporating organic chromophores into either side chains or polymers as discussed above. The underlying principle of this suppression is the resultant increase in the glass transition temperature of the particular system involved. Such an increase can also be attained via the formation of interchain chemical bonds resulting in the partial immobilization of organic chromophores in the resultant, cross-linked polymer matrices (Scheme 1(c)). Here, the desired overall effectiveness (i.e., in decreasing the mobility of organic chromophores in a polymer matrix) strongly depends on both the number and nature of cross-linkable sites made available. Not surprisingly, Marks and coworkers [9] have in fact reported improved thermal stabilities in a cross-linked epoxy guest–host system.

Along with the development of cross-linkable matrices, the three-dimensional structures of inorganic matrices have been considered as feasible matrices for the incorporation of organic NLO chromophores which could be bound covalently to matrices (Scheme 1(d)). Since these silica matrices possess excellent optical clarity, they should therefore maintain such clarity even after the incorporation of organic NLO chromophores. The high processing temperatures involved in performed inorganic matrices, however, presents a major obstacle in their subsequent adoption as matrices. Fortunately, this obstacle has been surmounted by the simple preparation of clear inorganic and modified inorganic–organic systems, both through the sol–gel process. The organic NLO chromophore may thus be incorporated into the matrices prepared by the sol–gel process in a similar fashion to that of polymer matrices. In fact, organic chromophores are versatile enough to be either physically or chemically incorporated into silica matrices. In these systems, the poled-order relaxation process has been dramatically prevented, accompanied with a high SHG macroscopic susceptibility after poling. A number of examples of organic NLO chromophores which have been incorporated into silica matrices via the sol–gel process have previously been reported [10–13].

Although significant advancements in thermal stability have recently been reported using sol–gel process, the preparation of materials with both higher thermal stability ($>150°C$), as well as SHG susceptibility, remains a topic worthy of future investigation.

In this work, the results of thermal stability studies along with the poled-order relaxation behavior and SHG intensities are reported for the chromophores [e.g., 2-methyl-4-nitroaniline(MNA, Scheme 2(a)), 3-nitro-9-ethyl-carbazole (carbazole 1, Scheme 2(b)), 3-nitro-9-hydroxyethyl-carbazole (carbazole 2, Scheme 2(c)), 4-ethyl (2-hydroxyethyl) amino-4′-nitro azobenzene (DR1, Scheme 2(d)) and modified DR1

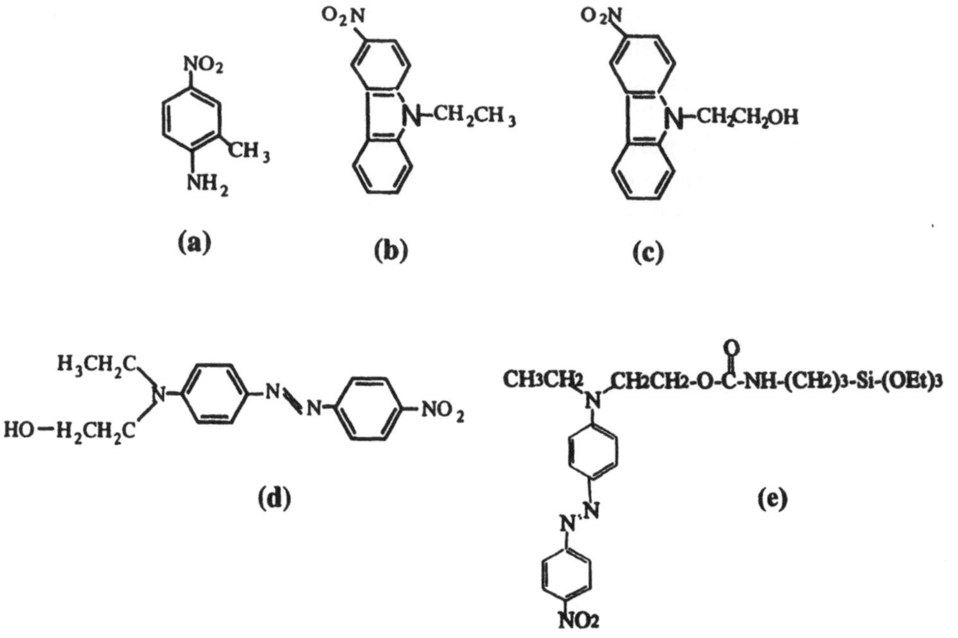

Scheme 2   Chemical structures of organic NLO chromophores.

with 3-isocyanatopropyl triethoxy silane (DRS, Scheme 2(e))] incorporated into silica matrices.

EXPERIMENTAL SETUP

Supply Materials Description

The materials MNA, carbazole 1, carbazole 2, and DR1 were purchased respectively from Aldrich and subsequently recrystallized as proper solvents before being used. DRS was prepared according to the methods outlined by Shul *et al.* [12]. Tetra-ethylorthosilicate (TEOS) was also purchased from Aldrich and used directly without further purification. The microscope glass slides used were then ultrasonically cleaned in a detergent solution, washed with deionized water, washed with isopropanol, and dried at 50°C dust free oven and stored in a desiccator. Two reagent-grade chemicals (i.e., concentrated HCl and ethanol) were both obtained and utilized as required. Double-distilled, deionized water was used in the all experiments.

Materials Preparation

The precursor solution for the silica formation consisted of TEOS, ethyl alcohol, HCl, water, and hexylene glycol (HG), the molar ratio for the precursor solution then being TEOS(1): EtOH(10): HG(0.1): HCl(0.05): H_2O(2): DMF(0.1). Next, organic NLO chromophores (10 wt% of TEOS) dissolved in DMF were added to the precursor solution, with the resultant mixtures then stirred for one hour. This solution yielded a viscosity of 3–5cp, thereby being suitable for spin-coating onto the glass substrate and in turn affording 1–2μm thickness thin composite films. During the preparation, these spin-coated thin films were prepared using 2000 rpm for 180 seconds and then further densified at 100°C for six hours.

Materials Characterization

A FT-IR was employed to monitor the gelation of a solution. A UV/VIS absorption spectra (Shimadzu 160A) are measured for the sample both before and after poling via a 15 kV corona discharge technique. During the thermal processing, both the polarized stability along with the subsequent degree of relaxation in the poled order of all spin-coated films was monitored by UV spectrometer. The change of viscosity with time was then measured by a concentric cylinder viscometer (Rotovisco RV3) equipped with an NV-ST-type of sensor. The second harmonic generation (SHG) intensity was investigated by a Nd: YAG Lager (10 mW/cm^2, pulse duration: ps) while the rotation angles of these samples were maintained in the range from −80° to 80°.

RESULTS AND DISCUSSION

Figure 1 represents the time-dependent viscosities for the silica matrices by varying molar ratio EtOH and TEOS employed in this work. The silica sol mixtures showed a slow increase in viscosity initially, but a subsequent sharp increase in value at the gel point (Figure 1). The possible coating-time ranges shown as dotted line. Based on these viscosity results, the optimum condition in the spin-coating process is therefore been as certained to be at a molar ratio of 10 (i.e., of EtOH to TEOS). At this molar ratio, the optical-quality, thin films of each sample were then conveniently prepared by spin-coating them directly onto the microscope glass slides [11].

As the sol–gel transformation of silica sol mixture progressed, the FT-IR spectra revealed the disappearance of two large bands at around 1100–1150 cm^{-1}, respectively, which would then suggest the underlying reason for the appearance of wide Si—O—C peaks of the ethoxy group attached to the TEOS while the Si—O—Si asymmetric stretching bonds around 1090–1160 cm^{-1} become increasingly stronger. Such results may thus indicate that Si—O—Si network structures are indeed forming [11].

UV spectra results were obtained for the organic NLO chromophores incorporated into the spin-coated silica films. The major spectral peaks appear to originate from the incorporated, organic NLO chromophores. Here, the maximum, MNA incorporated silica film peak appeared at 383 nm, with those of both carbazole 1 and carbazole 2 incorporated films appearing at 189 and 420 nm, respectively. In addition, both the DR1 and DRS incorporated films revealed maximum absorption peaks at 457 and 497 nm, respectively. It is therefore speculated in this work that the

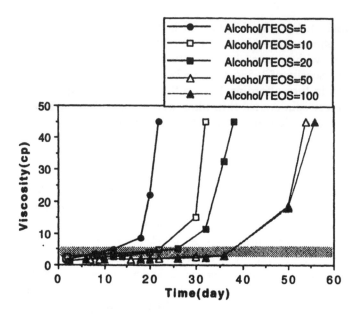

Figure 1   Viscosity changes of silica sol with aging time.

differences observed between the functional groups in carbazole 2 (i.e.,—C_2H_4OH) and carbazole 1, as well as those in DRS [—$CH_2CH_2CH_2$—Si—$(OR)_3$] and DR1 may account for the shifting observed in the maxima peaks of both carbazole 2 and DRS incorporated films [10–13].

The results obtained for the thermal degradation of organic NLO chromophores incorporated into silica films are shown in Figure 2. These samples were treated under ambient conditions at 100°C, where the degree of degradation of the organic NLO chromophores was determined by the absorbance change of maximum peaks. The decay rate of the absorbance peak in the silica films are on the order of: carbazole 1 > MNA > DR1 = carbazole 2 > DRS. These results thus indicate that the absorbance decay of both MNA and carbazole 1 incorporated silica films may be due to both the resultant sublimation and thermal degradation of the incorporated organic NLO chromophores. The silica network provides no reactive interaction to the organic NLO chromophores. In addition, the slight increase in thermal stability either of the carbazole 2 or DR1 incorporated silica films may in fact be the underlying reason for the chemical interaction observed between the hydroxy groups (i.e., in carbazole 2 and DR1) and the silica sol. The DRS incorporated silica films have revealed a more enhanced thermal stability than that of the other films analyzed in this work, due in particular to the numerous binding sites DRS possesses to induce a more effective incorporation with TEOS [10–13].

The fascinating poling phenomena attained for the DRS incorporated silica films are shown in Figure 3. To begin with, an UV absorbance change was observed after the film both poled and densified. The remarkable absorbance decrease provides indirect evidence for the alignment of incorporated organic NLO chromophores by

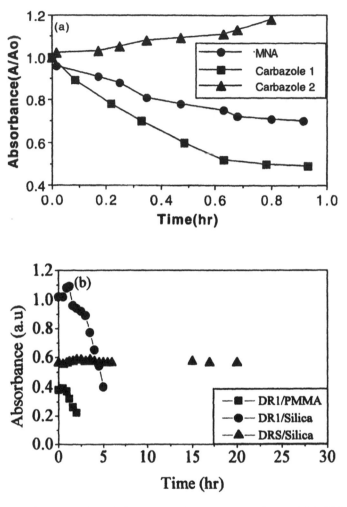

Figure 2 Thermal degradation of the organic NLO chromophores, (a) MNA, Carbazole 1, and Carbazole 2, (b) DR1 and DRS, incorporated silica films.

electric poling itself. Approximately 70% decrease in absorbance was observed from this work after poling of the DRS incorporated silica film [13].

In contrast to DRS incorporated silica films, however, the other silica films examined in this work yielded a slight recovery in their absorbances after 12 hours of electric field removal: this is due to the slow relaxation of the aligned carbazole 2 and DR1 residing in the silica matrices [11,13]. Figure 4 shows the UV absorbance changes of DRS incorporated film depending on the various temperatures. For the sample poled at 80°C shows a rapid decay of absorbance and this low absorbance maintained more than 200 hours. It means that ordered relaxation does not occur in DRS incorporated silica film. Based on these results, it then becomes evident that

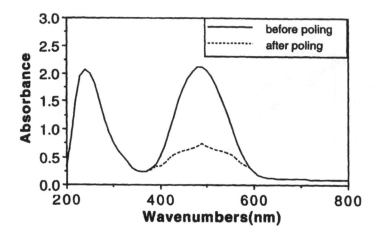

Figure 3   UV absorbance changes of DRS incorporated silica films before and after poling at 80°C.

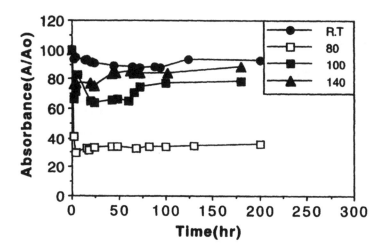

Figure 4   UV absorbance changes before and after poling, and relaxation behaviors of DRS incorporated silica films.

DRS may in fact be forming ordered arrays, with poling occurring at 80°C with these DRS arrays in turn becoming entrapped by the silica matrices. The optimum poling temperature implies that simultaneous alignment of DRS molecule and condensation reaction of Si—O—Si are important for the stabilization of aligned DRS molecule in silica. It is therefore speculated in this work that the incorporated DRS may be forming a liquid crystal phase at this temperature as well [14].

The SHG intensity for the corona-poled samples examined in this work was obtained with a Nd:YAG laser to observe the NLO properties. As seen in Figure 5, the SHG intensity of DRS in the silica matrix is large. Moreover, the SHG intensities

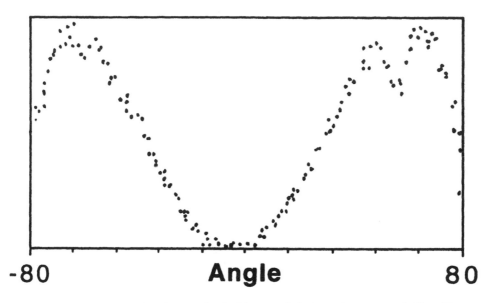

-80   **Angle**   **80**

Figure 5   SHG intensity changes for incident angle in DRS incorporated silica films.

of DRS change with the incident angle, as shown in Figure 5. The minimum value observed at this point was 0°, while the maximum was observed at 70°. This behavior is similar to that of DR1, albeit with slight deviation in the angle of maximum intensity. More detailed investigations on SHG intensity for DRS incorporate silica film involving the variation of both poling temperatures and initial DRS concentrations remain a topic of current investigation [12].

CONCLUSION

The thermal stability of MNA, carbazole 1, carbazole 2, DR1 and DRS have been investigated using UV/VIS absorption spectra at 100°C. Among them, the DRS incorporated silica films yield the most thermally stable matrices. From the results of the UV studies, it is concluded that temperature plays an important factor along with NLO chromophore molecular structure in the resultant orientation of NLO molecules in the matrices themselves. In addition, the orientation of the NLO chromophores arrays can be effectively hindered from further rotation by their chemical incorporation into the silica matrices.

The result of NLO stability demonstrate the possibility of the efficient design process for the stabilization of macroscopic second-order optical nonlinearities using reactive sol–gel transformation.

J.-S. LEE *et al.*

ACKNOWLEDGMENTS

The authors gratefully extend their thanks to Dr. M. G. Choi at Yonsei Univ., Korea. We also acknowledge to Dr. H. Sasabe and T. Wada at RIKEN, Japan for helpful discussions and the SHG intensity measurements provided.

REFERENCES

1. Ulich, D. (1990). *Mol. Cryst. Liq. Cryst.* **11**, 366.
2. Parasad, P. N. and Williams, D. J. (1991). *Introduction to Nonlinear Optical Effects in Molecules and Polymers*, John Wiley & Sons, New York.
3. Carter, G. M. (1987). *J. Opt. Soc. Am. D* **4** 1018.
4. Lin, J. T. (1987). *Laser Focus* **23**, 59.
5. Singer, K., Sohn, J. and Lalama, S. (1986). *Appl. Phys. Lett.* **49**, 248.
6. Xu. C., Wu, B., Dalton, L. R., Shi, Y., Ranon, P. M. and Steier, W. H. (1992). *Macromolecules* **25**, 6714, and references therein.
7. Kim, J., Plawsky, J. L., LaPeruta, R. and Korenowski, G. M. (1992). *Chem. Mater.* **4**, 249.
8. Kim, J.-J., Zyung, T. and Hwang, W.-Y. (1994). *Appl. Phys. Lett.* **64**, 3488.
9. Hubbard, M. A., Mark, T. J., Yang, J. and Wong, G. K. (1989). *Chem. Mater.* **1**, 167.
10. Shul, Y. G., Lee, D. J., Lee, J.-S. and Wada, T. J. (1993). *Ceramic Soc. Japan* **101**, 76.
11. Jung, M. W., Shul, Y. G., Mun, J. H. and Wada, T. (1994). *Mol. Cryst. Liq. Cryst.* **247**, 111.
12. Shul, Y. G., Jung, K. T., Cho, G., Lee, J.-S., Wada, T., Sasabe, H., Jung, M. W. and Choi, M.-G. (1996) *Nonlinear Optics*, **15**, 411.
13. Lee, J.-S., Cho, Y.-S., Cho, G., Jung, K. T., Shul, Y. G. and Choi. M.-G., (1996) *Mol. Cryst. Liq. Cryst.*, **280**, 53.
14. Abe, J., Hasegawa, M., Matsushima, H., Shirai, Y., Nemoto, N., Naguse, Y. and Takamiya, N. (1995). *Macromolecules* **28**, 2938.

5a. EFFECTS OF DRAWING AND POLING ON NONLINEAR OPTICAL COEFFICIENTS OF POLED POLYMERS

TOSHIYUKI WATANABE, JONG-CHUL KIM, and SEIZO MIYATA

Graduate School of Bio-applications and Systems Engineering,
Tokyo University of Agriculture and Technology

1 INTRODUCTION

Organic molecular and polymeric materials that display interesting nonlinear optical properties have received a great deal of attention for optical communication and integrated optics [1–3]. A very wide variety of organic materials that include molecular crystals [4], composites [5–8], liquid crystals [9], nonlinear optics (NLO)-dye functionalized polymers [10–21] have been investigated for nonlinear optics. Of particular interest when addressing the issue of second harmonic generation (SHG), have been concentrating on one-dimensional (1-D) charge transfer (CT) molecules which exhibit very large molecular hyperpolarizabilities arising from the extended π-conjugation terminated by strong donor and acceptor groups [22]. 1-D CT molecules are most often employed either as a component of a composite (Guest-host) system or NLO active part embedded into a polymer chain. Recently two-dimensional (2-D) charge transfer molecules [23,24] and octupolar molecules [25] were synthesized and their second-order optical nonlinearity were investigated. 2-D CT molecules possess the large off-diagonal tensor components β_{xyy} rather than diagonal component. The remaining nonvanishing irreducible β_{xyy} component has not been previously considered in the perspective of molecular engineering and optimization as proposed in this work. We have found that 2-D CT molecules have a great potential for frequency doubling.

In this monograph, we demonstrate effects of drawing of poled polymers on nonlinear optical coefficients. We synthesize the novel phenoxy polymers containing *p*-nitrophenylcarbamate as a side chain. The chromophore ratio, side chain having large molecular polarizability α and dipole moment is varied to confirm the effect of chromophore ratio on SHG activity. After the characterization of materials, uniaxially and biaxially drawing of side chain polymers were examined and order parameters were obtained from these experiments. The effects of drawing of main chain polymer also investigated.

In addition, we discuss the origin of, β_{xyy} components and contribution of off-diagonal tensor components of molecular hyperpolarizability. Furthermore, the advantage of 2-D CT molecules for the practical applications of SHG is demonstrated.

2 THEORY

In this section let us consider relationship between nonlinear optical properties of poled and drawn glassy polymers and molecular hyperpolarizability. The relationship between molecular frame and laboratory frame were illustrated by Figure 1, where X, Y and Z represent the laboratory frame, respectively. X is parallel to the drawing direction and Z represents the polar axis induced by poling. The x, y and z axes are the molecular frame. Euler angles, θ, φ and α, are introduced to transform the laboratory frame to molecular frame. The dipole moment of molecule is parallel to the x axis. In the dipole approximation and for instantaneous response, the

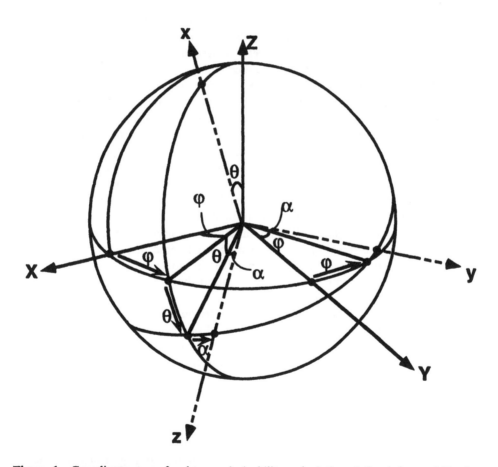

Figure 1 Coordinate axes for hyperpolarizability calculation defined for a 2-D charge transfer molecular system. The X, Y and Z are the poled polymer axes and x, y and z are the molecular axes. The θ defines the tilt angle to the film normal and φ defines the azimuthal angle which represents the angle between Z and projection of the x to 2–3 plane.

polarization P is defined as the dipole moment per unit volume and may be expressed as a power series in the electric field E.

$$P_I = \chi_I^{(0)} + \chi_{IJ}^{(1)} E_J + \chi_{IJK}^{(2)} E_J E_K + \cdots, \qquad (1)$$

where the tensor quantity $\chi^{(n)}$ is the nth order susceptibility. A centrosymmetric bulk material will not exhibit any SH light. The nonlinear optical coefficients depend on both degrees of alignment and magnitude of the molecular hyperpolarizability.

The microscopic polarization p_i is similar to Eq. (1) and given by

$$p_i = \mu_i + \alpha_{ij} F_j + \beta_{ijk} F_j F_k + \cdots, \qquad (2)$$

where F_m is the local electric field, μ is the molecular ground-state dipole moment, α_{ij} is the linear polarizability and β_{ijk} is the first-order hyperpolarizability. If the interaction between each molecule can be neglected, the macroscopic nonlinearity, b_{IJK} was expressed as the summation of microscopic nonlinearity β by considering the thermodynamic average of p_i over all possible orientation. The macroscopic polarizability, b_{IJK}, is expressed by

$$b_{IJK} = \int_0^{2\pi} d\Omega \, a_{Ii}^{-1} a_{Jj} a_{Kk} G(\Omega, E_p) \beta_{ijk}, \qquad (3)$$

where a_{IJ} are the rotation matrices between the molecular frame, denoted by lower-case subscript and laboratory frame, denoted by upper-case subscript. a_{IJ}^{-1} is the inverse matrices of a_{IJ}. The function $G(\Omega, E_p)$ can be expanded in terms of Legendre polynomials,

$$G(\theta, E_p) = \sum_{l=0}^{\infty} \frac{2l+1}{2} A_l P_l(\cos \theta) \qquad (4)$$

with

$$A_l = \int_{-1}^{1} d(\cos \theta) G(\theta, E_p) P_l(\cos \theta). \qquad (5)$$

The A_l are the ensemble average of the P_l, $\langle P_l \rangle$ and are defined as the macroscopic order parameters associated with an axially symmetric liquid crystal.

The nonlinear optical coefficient is expressed in terms of b_{IJK}

$$d_{IJK} = N f_I^{2\omega} f_J^{\omega} f_K^{\omega} b_{IJK}, \qquad (6)$$

where f is the local field factor.

Finally the following equation can be derived from the combination of Eqs. (1)–(3). Only β_{xxx}, β_{xyy} and β_{xzz} components affect nonlinear optical coefficients and other tensor components were canceled out during a poling process. We assume that the point group of poled and drawn polymer films belongs to mm2 as same as Poly(vinylidene fluoride) (PVDF). We would expect three independent tensor components because of lacking the cylindrical symmetry.

$$b_{33} = L_3(p)\beta_{xxx} + 3\langle\cos^2\alpha\rangle(L_1(p) - L_3(p))\beta_{xyy} + 3\langle\sin^2\alpha\rangle(L_1(p) - L_3(p))\beta_{xzz}, \quad (7)$$

$$\begin{aligned} b_{31} = &\langle\cos^2\varphi\rangle(L_1(p) - L_3(p))\beta_{xxx} + (\langle\cos^2\alpha\rangle\langle\cos^2\varphi\rangle L_3(p) \\ &+ \langle\sin^2\alpha\rangle\langle\sin^2\varphi\rangle L_1(p))\beta_{xyy} - 2(\langle\cos^2\alpha\rangle\langle\cos^2\varphi\rangle(L_1(p) \\ &- L_3(p)))\beta_{xyy} + (\langle\sin^2\alpha\rangle\langle\cos^2\varphi\rangle L_3(p) \\ &+ \langle\cos^2\alpha\rangle\langle\sin^2\varphi\rangle L_1(p))\beta_{xzz} \\ &- 2(\langle\sin^2\alpha\rangle\langle\cos^2\varphi\rangle(L_1(p) - L_3(p)))\beta_{xzz}, \end{aligned} \quad (8)$$

$$\begin{aligned} b_{32} = &\langle\sin^2\varphi\rangle(L_1(p) - L_3(p))\beta_{xxx} \\ &+ (\langle\cos^2\alpha\rangle\langle\sin^2\varphi\rangle L_3(p) + \langle\sin^2\alpha\rangle\langle\cos^2\varphi\rangle L_1(p))\beta_{xyy} \\ &- 2(\langle\cos^2\alpha\rangle\langle\sin^2\varphi\rangle(L_1(p) - L_3(p)))\beta_{xyy} \\ &+ (\langle\sin^2\alpha\rangle\langle\sin^2\varphi\rangle L_3(p) + \langle\cos^2\alpha\rangle\langle\cos^2\varphi\rangle L_1(p))\beta_{xzz} \\ &- 2(\langle\sin^2\alpha\rangle\langle\sin^2\varphi\rangle(L_1(p) - L_3(p)))\beta_{xzz}, \end{aligned} \quad (9)$$

where $L_1(p)$ and $L_3(p)$ are first and third order Langevin function, respectively. The poling field p is given by

$$p = \frac{\mu E}{kT}, \quad (10)$$

where μ, E and k represent the dipole moment, poling electric field and Boltzman constant, respectively. When p is smaller than 1 and the film was not drawn, Eqs. (7) and (8) give the same results obtained by EFISH measurement. In the case of one-dimensional charge transfer molecules, order parameter $\langle P_1\rangle$ and $\langle P_3\rangle$ can be introduced to explain the degree of orientation.

$$d_{33} = N\beta_{zzz}(\tfrac{3}{5}\langle P_1\rangle + \tfrac{2}{5}\langle P_3\rangle), \quad (11)$$

$$d_{31} = N\beta_{zzz}(\tfrac{1}{5}\langle P_1\rangle + \tfrac{1}{5}\langle P_3\rangle). \quad (12)$$

Second harmonic generation therefore measures $\langle P_1\rangle$ and $\langle P_3\rangle$ directly.

3 EFFECTS OF DRAWING OF SIDE CHAIN POLYMER ON NLO COEFFICIENTS

The novel side chain polymers were synthesized as shown in Figure 2. Phenoxy resin (Tohto Kasei Co. LTD.) having average molecular weight (Mw) of 75,000 was used for polymer reaction, since we need a free standing film for drawing. Phenoxy resin was dissolved in THF and purified by precipitation in methanol. p-nitro-phenylisocyanate was reacted with phenoxy resin with di-n-butyl-tindilaurate. The polymer was purified by precipitation in methanol three times after dissolving in THF, and dried in a vacuum oven at 100°C for two days. In the same procedure, the mole ratio of p-nitrophenylisocyanate against 1 mole of phenoxy resin varied from 0.4 mole to 1.1 mole. All of these polymers were slightly yellow powder and could easily dissolve in THF, 1,4-dioxane, and dimethylformamide (DMF).

In order to investigate the optimum poling condition, spin-coated films with 1 μm thickness were prepared on a soda lime glass from 10 wt% 1,4-dioxane solution. These films were dried in a vacuum for 3 days above their glass transition temperature. The thickness and refractive indices of the films were determined by DEKTAK IIA and prism-coupling method [26].

High positive voltage in the range of 4–9 KV was applied to the spin-coated films by corona poling technique wherein the evaporated aluminum layer on the other side was used as the ground electrode. Poling temperature was optimized during *in-situ* second harmonic generation measurement. Chromophore orientation of polymers was analyzed by UV/Visible spectrophotometer before and after poling.

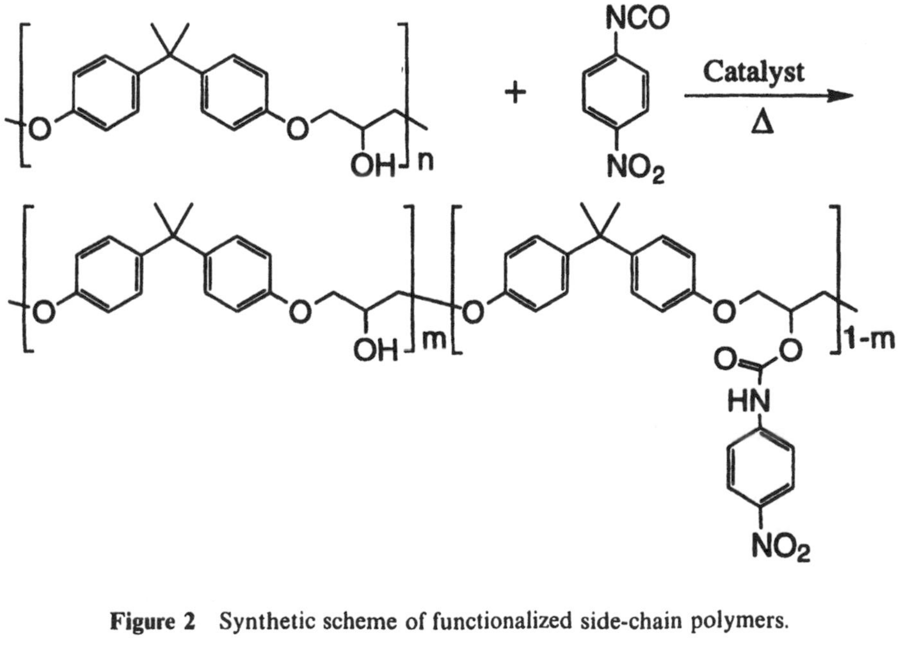

Figure 2 Synthetic scheme of functionalized side-chain polymers.

The films with about 70 μm thickness were prepared by cast from 25 wt% 1,4-dioxane solution. These solvent-cast films were dried in a vacuum oven for 10 days with increasing temperature. The obtained films were dried above T_g temperature. Finally these isotropic films were uniaxially and biaxially drawn at near T_g temperature. Draw ratio, here in defined as the ratio of the final length of the drawn film in the stretching direction relative to the length before stretching, was varied from 1.3 to 2.3. The thicker pristine and drawn polymer films were poled by contact poling method. The evaporated aluminum layers on both sides of the film were used as electrodes. These films were applied by 75–150 V/μm positive electric field at room temperature, and the temperature was increased by $T_g - 17°C$ and maintained for 2 hours. The film is then cooled to room temperature with the electric field still applied and the electric field was finally removed. The refractive indices were measured by the same instrument.

The chemical structure of polymers was characterized by ^{1}H-NMR and FT-IR. The ratio of 4-nitrophenylcarbamate substituted in phenoxy resin, namely chromophore ratio, was varied from 28% to about 100%. Glass transition temperature (T_g) of these polymers was increased up to 130°C with the increasing chromophore ratio because of hydrogen bonding formation, while T_g temperature of the polymer substituted by side chain such as benzoyl group was decreased with the increasing the esterification ratio. This phenomenon of T_g temperature increases due to hydrogen bonding formation also showed in phenoxy polymer containing α-cyano unsaturated carboxylate. Cut-off wavelength of the synthesized polymer was largely shifted compared with phenoxy polymer due to the introduction of nitro group. As the result, the color of the synthesized polymer film having about 100% chromophore ratio was yellowish. Before decomposition, no melting peak was observed in these polymers. No sharp peaks were observed by X-ray diffraction, indicating that they are amorphous.

To find optimum poling temperature, *in situ* SHG measurements were carried out as shown in Figure 3. According to the increasing temperature, chromophore of the polymer chains are aligned parallel to the direction of applied electric field, so that noncentrosymmetric structure was formed by poling. The SHG signal is proportional to square root of poling electric field and the SHG signal would typically show the maxima at certain temperature and then decrease due to the ionic conductivity of polymer film. Thus the optimum poling temperature of the spin coated films having different chromophore ratio was increased with the increasing chromophore ratio, but the temperatures were almost 17°C lower than glass transition temperature.

The absorption spectra of polymer films before and after corona poling were measured with UV/Visible spectrophotometer in the wavelength range of 200–800 nm as shown in Figure 4. After poling, a decrease in absorbance (hypochromic shift) was observed in all polymers, and the decreasing ratio was increased with the increasing chromophore ratio. The decrease in absorbance after poling is due to the alignment of chromophore. During corona poling, surface charge is accumulated on the surface of the polymer films with opposite charge at the planar electrode [27], and this introduces a large electrostatic field in the film that interacts with the chromophores of the polymer. The static field aligns the dipoles in the direction of

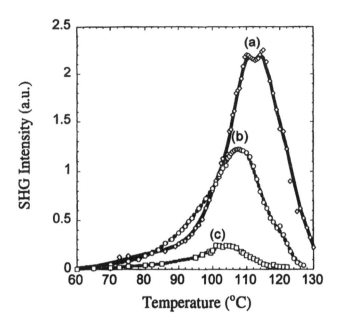

Figure 3 *In situ* SHG measurements of the polymer films having chromophore ratio (1-m), (a) 0.99, (b) 0.72, and (c) 0.52, respectively.

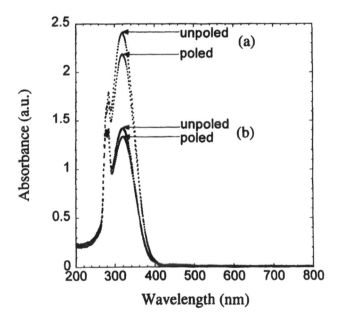

Figure 4 UV/Visible absorption spectra of the polymers before and after poling and the corresponding chromophore ratio (1-m), (a) 0.99 and (b) 0.52, respectively.

the poling field, which leads to a change in the intensity of the absorption spectrum, i.e., to dichroism. Order parameter to express the degree of poling-induced orientation is given by

$$\Theta = 1 - A_p/A_o, \tag{13}$$

where A_p is the absorbance perpendicular to the poling direction of a poled film and A_o is the absorbance of an unpoled polymer film. The order parameter of spin-coated film was determined to be 0.10 at about 0.99 chromophore ratio and 0.05 at about 0.54 chromophore ratio and it was also increased with the increasing chromophore ratio.

Both drawn and poled polymer films were studied to examine the linear and nonlinear optical properties. All prepared films showed the isotropic relation of refractive index such as $n_X = n_Y = n_Z$ because the polymer chains were randomly oriented by drying above T_g temperature for 2 days. The isotropic polymer films having 0.99 chromophore ratio were poled by contact poling method in 75 and 150 V/μm. And the same isotropic polymer films were uniaxially drawn about 1.8 times and poled on the same poling condition. Using these films, refractive indices were measured and nonlinear optical coefficients were measured by Maker fringe method. The configuration of SHG measurement is shown in Figure 5. Q-switched Nd:YAG laser (1064 nm, 10 ns/pulse, 10 Hz) was used for Maker fringe measurements. The second harmonic signal was obtained in the transmission and selected from a monochromator. It was detected by photomultiplier tube and averaged by a boxcar integrator. The Maker fringe data were analyzed by fitting parameters to the appropriate theoretical formula [15]. Maker fringes for both *s*- and *p*-polarized fundamental and *p*-polarized second harmonic of pristine and poled polymer films are measured. The amorphous poled polymer belongs to mm∞ point group, and is changed to mm2 point group by drawing. The symmetry operations include a two-fold axis along the film normal direction and two mirror planes perpendiculars to each other and including the 2-fold axis.

For a system with mm2 point group, the nonlinear polarization P_{NL} is given by

$$P_{NL} = \begin{bmatrix} 0 & 0 & 0 & 0 & d_{15} & 0 \\ 0 & 0 & 0 & d_{24} & 0 & 0 \\ d_{31} & d_{32} & d_{33} & 0 & 0 & 0 \end{bmatrix} \begin{bmatrix} E_1^2 \\ E_2^2 \\ E_3^2 \\ 2E_2E_3 \\ 2E_1E_3 \\ 2E_1E_2 \end{bmatrix}, \tag{14}$$

five nonzero d coefficients are existed. Under the Kleinman symmetry restriction, we have

$$d_{15} = d_{31}, \qquad d_{24} = d_{32}.$$

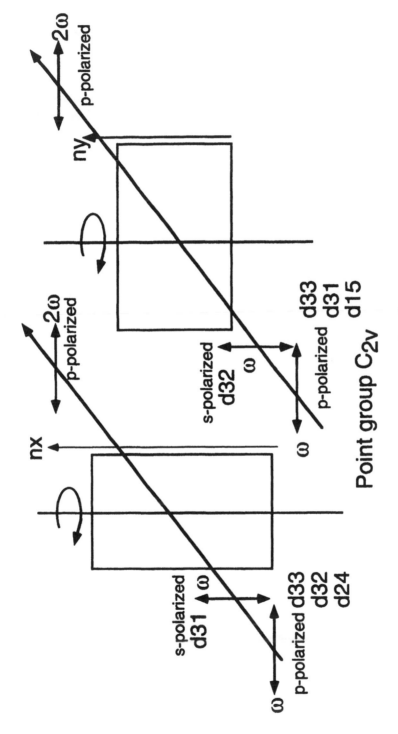

Figure 5 Configuration of SHG measurement.

The second-order NLO properties are finally governed by the three independent nonzero coefficients, if there is no absorption. Accordingly two independent nonlinear optical coefficients, d_{33} and d_{31}, can be calculated in as prepared poled polymer because the polymer chains are randomly oriented. In drawn and poled polymer five independent nonlinear optical coefficients, d_{33}, d_{31}, d_{32}, d_{24}, and d_{15}, are calculated due to orientation of polymer chain. Thus Maker fringes for both *s*- and *p*-polarized fundamental and *p*-polarized second harmonic of uniaxially drawn and poled polymer films are measured to the drawing direction and perpendicular direction against drawing direction, and an example for which is shown in Figure 6. Here Φ is the angle between the propagation direction and the normal to the film surface. The SHG intensity is zero at 0° and grows symmetrically around this angle, indicating that after poling the average dipole moment of the polymer is perpendicular to the film surface. The nonlinear optical coefficients can be estimated by fitting the measured SHG intensities in both *p–p* and *s–p* polarization compared with *Y*-cut quartz reference and the calculated results are shown in Table 1.

The d_{33} value of the pristine and poled polymer film was increased from 2.4 pm/V to 4.0 pm/V with the increasing applied electric voltage. By drawing, this d_{33} value was increased from 4.0 to 6.4 pm/V with the increasing applied electric voltage. And d_{33} values of as prepared films were about three times larger than that of as d_{31}, whereas one film drawn and poled by 150 V/μm showed about five times as large as d_{31}. The ratio d_{31}/d_{32} of all poled polymer films were found to be 1 and not changed by drawing. These results suggest that the side chain could be randomly oriented against in-plane of polymer. Figure 7 shows the Maker fringes for both *s*- and

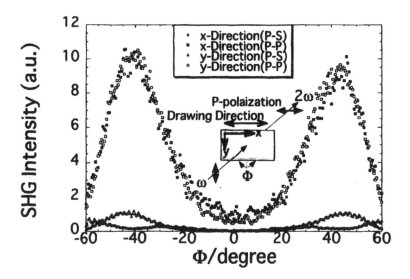

Figure 6 Maker fringe of uniaxially drawn and poled polymer film with *p–p* and *p–s* polarization at the drawing direction and its perpendicular direction; chromophore ratio of polymer film with 46 μm was 0.99; draw ratio is 1.8 and applied electric field was 150 V/μm.

Table 1 Nonlinear optical coefficients and order parameter of poled polymers.

Sample	d_{33} (pm/V)	d_{31} (pm/V)	d_{32} (pm/V)	a	$\langle P_1 \rangle$	$\langle P_2 \rangle$
as-prepared film	4.0	1.1	1.1	0.28	0.54	0.06
uniaxially drawn film	6.4	1.4	1.2	0.20	0.62	0.17
biaxially drawn film	8.8	1.3	1.4	0.15	0.98	0.41
under stress*	—	—	—	0.70	0.041	−0.02
without under stress*	—	—	—	0.33	0.16	0.00

*Data from M.G. Kuzyk, J. Opt. Soc. Am. B., **6**, 742 (1989).

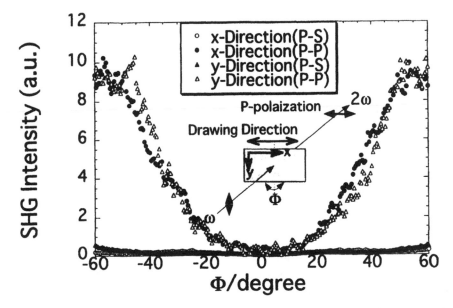

Figure 7 Maker fringe of biaxially drawn and poled polymer film with p–p and p–s polarization at the drawing direction and its perpendicular direction; chromophore ratio of polymer film with 34 μm was 0.99; draw ratio is 1.4 and applied electric field was 150 V/μm.

p-polarized fundamental and p-polarized second harmonic of biaxially drawn and poled polymer films. The anisotropy ratio, d_{31}/d_{33} became larger compared with uniaxially drawing. The ratio d_{31}/d_{32} of all films were also 1. The order parameter $\langle P_1 \rangle$ and $\langle P_3 \rangle$ was evaluated using oriented gas description. The calculated results are shown in Table 1. Here we define the anisotropic factor a,

$$a = d_{31}/d_{33}.$$

$\langle P_1 \rangle$ and $\langle P_3 \rangle$ of uniaxially drawn and biaxially drawn films are extremely large compared with the films poled under stress [28]. The out-of-plane birefringence of drawn and poled polymer films were larger than that of as prepared poled polymer films. The uniaxially drawn films showed the relation of refractive index such as $n_x > n_y = n_z$. Biaxially drawn and poled polymer films show larger out-of-plane birefringence (Δn_{z-y}) than that of uniaxially drawn films. This large out-of-plane birefringence of drawn films suggests that NLO chromophore tend to align normal to the film surface. These results reveal that preferential orientation of NLO chromophore induced by drawing give the large order parameter.

4 EFFECTS OF DRAWING OF MAIN CHAIN POLYMER ON NLO COEFFICIENTS

The main chain polymer consists of 2-D CT molecules was synthesized [29,30] to confirm the effects of drawing. The chemical structure used in this experiments is shown in Formula 1. The polymer abbreviated as U-1 was dissolved in N-methyl-2-pyrrolidone (NMP) to form solutions of various concentrations. The amount of NMP was adjusted to give rise to a desired viscosity suitable for casting. The solution was filtered using a membrane filter with 1 μm pores to remove the dust and undissolved particles. Films were prepared by spin coating on the soda-lime glass. After drying the film at 140°C in vacuum oven for 24 hours the glass transition temperature of polymer film were measured by differential scanning calorimetry (DSC). The glass transition temperature of U-1 was found to be 175°C. The refractive indices of the samples were determined by a prism coupling method using the different laser sources. The prism coupler was operated in accordance with the optical waveguide principle with the polymer film served as the propagation layer in the slab waveguide configuration [26].

The aluminium electrode was deposited on the backside of the soda-lime glass for poling. The polymer films were poled just below the glass transition temperature for 2 hours by corona poling configuration [27].

Nonlinear optical coefficients were measured according to the Maker fringe method [31] using Q-switched pulse Nd:YAG laser (1064 nm, 10 ns) as mentioned in Section 3. The molecular hyperpolarizabilities were theoretically calculated by MOPAC-PM3 method.

The relationship between hyperpolarizability and nonlinear optical coefficients were represented by the oriented gas description [22]. The measured nonlinear optical coefficients were compared with the theoretical value. The β of polyurea was obtained by using of model compounds in Figure 8. The MOPAC-PM3 calculation

Formula 1   Chemical structure of polymer containing 2-D CT molecules.

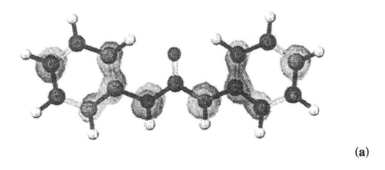

(a)

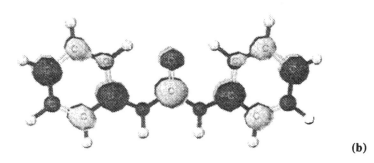

(b)

Figure 8   Molecular orbitals of model compound of polyurea. (a): HOMO, (b): LUMO.

gives only the off-resonant β_x^0 value in accordance with finite-field method. To account for the dispersion, a two-level quantum system is used to describe the dispersion of β. The β can be presented as follows:

$$\beta_x^{2\omega} = F(W, \omega) \times \beta_x^0. \tag{15}$$

The dispersion factor is given by

$$F(W, \omega) = 1/\{[1 - (2h\omega/W)^2][1 - (h\omega/W)^2]\}, \tag{16}$$

where W is band gap of molecule and ω is angular frequency. The calculation results is summarized in Table 2.

   The molecular orbital of model compound is shown in Figure 8, where the size of circles for each atom is proportional to the coefficient of p_π basis function. The HOMO (Figure 8(a)) is predominantly of nitrogen p_π character, while the LUMO (Figure 8(b)) includes significant contribution from the carbonyl group. Thus the excitation from HOMO to LUMO excitation should correspond to two nitrogen to the carbonyl oxygen transition. The two directional charge transfer was observed in

Table 2 First-order hyperpolarizability of model compound calculated by MOPAC-PM3.

Component	$(\omega = 0)$ $\times 10^{-30}$ esu	$(\omega = 1.17\,\text{eV})$ $\times 10^{-30}$ esu
β_{xxx}	0.27	0.40
β_{xyy}	-3.20	-4.77
β_{xzz}	0.15	0.22

this system which is the origin of large β_{xyy} components. The β_{xyy} component of model compound dominates almost x component of β tensor.

The nonlinear optical coefficients of poled polymer were measured as a function of draw ratio. The polymer films were uniaxially drawn at 180°C. It is well known that d_{33} of poled polymers without drawing is about 3 times larger than d_{31}. The d_{31}/d_{32} ratio of side chain polymers containing 1-D CT molecule is independent on the drawn ratio. Whereas d_{31} of main chain polymer containing 2-D CT molecule increases with increase of draw ratio. If it is assumed that Euler angle θ and ϕ become zero by drawing, then Eqs. (8) and (9) can be simplified and given by

$$b_{33} = \frac{1}{15}\frac{\mu E}{kT}(3\beta_{xxx} + 6\beta_{xyy}), \tag{17}$$

$$b_{31} = \frac{1}{15}\frac{\mu E}{kT}(2\beta_{xxx} - \beta_{xyy}). \tag{18}$$

In order to obtain the efficient SHG by use of birefringence phase-matching, large d_{31} is required. The d_{31} value of main chain polymers consisting of 2-D CT molecules can be controlled by drawing. It must be emphasized that the sign of β_{xxx} and β_{xyy} should be different to optimized b_{31}. Off-diagonal tensor components of NLO chromophore of polyurea can be enhanced by drawing, since sign of β_{xxx} and β_{xyy} is different. The d constant is plotted as a function of draw ratio in Figure 9.

The d_{31} of polyurea increases with increasing draw ratio, whereas d_{33} decreases. This result is consistent with Eqs. (9) and (10) we derived. The $\langle \cos^2 \alpha \rangle$ and $\langle \cos^2 \varphi \rangle$ were evaluated from d_{33}, d_{31} and d_{32} using Eqs. (7)–(10) as shown in Figure 10. Both parameters $\langle \cos^2 \alpha \rangle$ and $\langle \cos^2 \varphi \rangle$ increase with increasing of draw ratio. It seems that there is no dependence of the ratio of $\langle \cos^2 \alpha \rangle / \langle \cos^2 \varphi \rangle$ on draw ratio. The obtained film (draw ratio is 1.28) is still far from the perfect orientation of main chain. When the main chain of polymer is completely aligned, the sign of d_{33} and d_{31} will be different.

In order to control the phase-matching condition using birefringence of polymer the refractive index of drawing direction should be larger than that of poling direction. Figure 11 shows the birefringence of side chain and main chain polymer after poling.

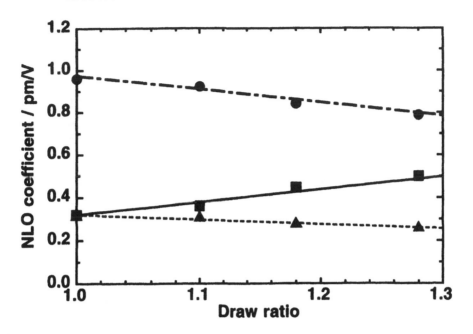

Figure 9   NLO coefficients of polyurea as a function of draw ratio. Filled circle: d_{33}, Filled square: d_{31}, Filled triangle: d_{32}.

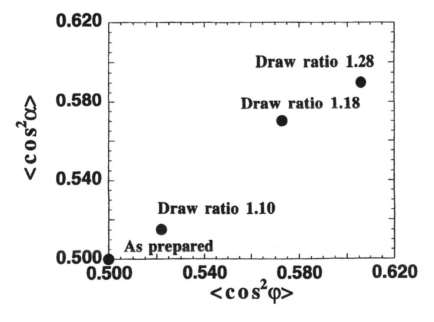

Figure 10   The plots of relationship between $\langle \cos^2 \alpha \rangle$ and $\langle \cos^2 \varphi \rangle$.

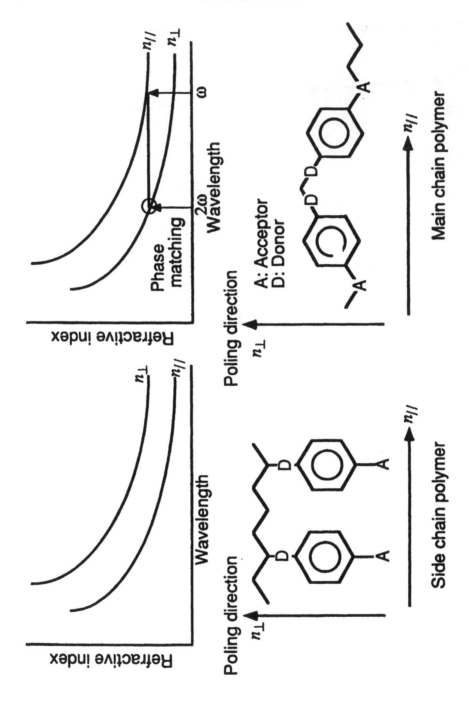

Figure 11 The schematic figure of dispersion of refractive indices.

There is no phase-matching point of side chain polymer, since $n_\perp$ is always larger than $n_{//}$. However, one can find the phase-matching point in main chain polymer due to the preferential orientation of mesogenic groups [29,30]. In fact, polyurea shows positive birefringence as shown in Figure 11.

In this point of view main chain polymer consisting of 2-D CT molecules can optimize both dispersion of refractive indices and off-diagonal tensor components for phase-matching by drawing and poling. The phase-matching experiments were performed using the polyurea [32]. The result of angle turning type-I phase-matched SHG at room temperature is shown in Figure 12.

The calculated phase-matching direction at y–z plane was confirmed by the experiment indicating our refractive indices measurements have good accuracy. This phase-matching is critical. In the case of main chain type polymers, the noncritical phase-matching is also possible by controlling the drawing ratio. The experiment for noncritical phase-matching is under going. The phase-matching conditions at different draw ratio were investigated. The type I phase-matching loci of polyurea was shown in Figure 13. The polar angle θ decreased with increasing draw ratio. At certain draw ratio noncritical phase-matching may be realized.

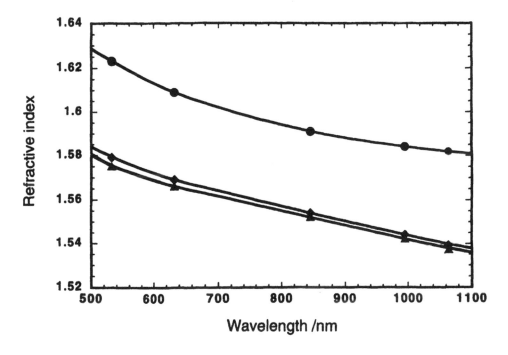

Figure 12 Dispersion of refractive indices of polyurea (draw ratio: 1.28). Filled circle: nx, Filled square: ny, Filled triangle: nz.

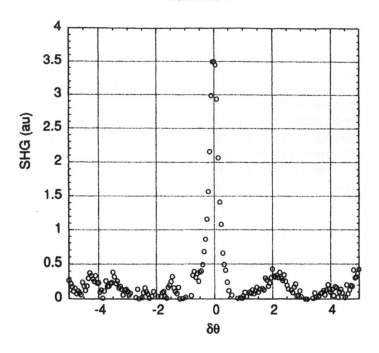

Figure 13   Phase-matched SHG of polyurea as a function of incident angle.

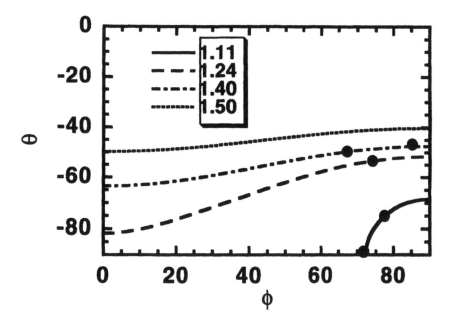

Figure 14   Phase-matching loci of polyurea. Filled circles: experimental data; Lines: theoretically calculated curves.

5 SUMMARY

A method for controlling the ratio of nonlinear optical coefficients a with drawing was demonstrated. In the case of side chain polymer, the nonlinear optical properties d_{33} was increased with increasing of draw ratio. The order parameter of biaxially drawn side chain polymer was found to be very high due to the preferential orientation of NLO chromophore by stretching. Our studies on polyurea have shown that 2-D CT molecules are novel NLO materials in which off-diagonal β components play an important role in optimizing d_{31} constant after drawing. One of advantages offers by 2-D CT molecules embedded into main chain polymer over side chain polymer is their ability to satisfy the phase-matching condition in a bulk media. Studies on analogous 2-D CT molecule should stimulate further research.

ACKNOWLEDGMENTS

The author wish to thank Dr. X. T. Tao, Tokyo University of Agriculture & Technology, Dr. H. S. Nalwa, Hitachi Co. Ltd., and Dr. Lee, Kum-oh National University of Technology, for useful discussions. This research was partially supported by Grant-in-Aid Research #07246104, #07246105 and #07750059.

APPENDIX

The nonlinear optical coefficients d_{IJK} can be expressed in terms of the molecular second-order susceptibility b_{IJK} as

$$d_{IJK} = N f_I^{2\omega} f_J^{\omega} f_K^{\omega} b_{IJK}, \tag{A1}$$

where N is the number density and b_{IJK} is an orientational average of second-order susceptibility. The indices I, J and K define the coordinate system of the molecule, while indices i, j and k define those of the molecular axis. The two coordinate systems are related by Euler angles, θ, α and φ.

The orientational average in the presence of a poling field E_p is of the form

$$b_{IJK} = \int_0^{2\pi} d\Omega\, a_{Ii}^{-1} a_{Jj} a_{Kk} G(\Omega, E_p) \beta_{ijk}, \tag{A2}$$

where the volume element is given by

$$\int d\Omega = \int_0^{2\pi} d\varphi \int_0^{2\pi} d\alpha \int_{-1}^{1} d(\cos\theta), \tag{A3}$$

a_{IJ} is the transformation matrix:

$$a_{IJ} = \begin{pmatrix} a_{11} & a_{12} & a_{13} \\ a_{21} & a_{22} & a_{23} \\ a_{31} & a_{32} & a_{33} \end{pmatrix}, \tag{A4}$$

where each components of a_{ij} is given by

$$a_{11} = \cos\alpha\cos\varphi\cos\theta - \sin\alpha\sin\varphi, \tag{A5}$$

$$a_{12} = \cos\alpha\sin\varphi\cos\theta + \sin\alpha\cos\varphi, \tag{A6}$$

$$a_{13} = -\cos\alpha\sin\theta, \tag{A7}$$

$$a_{21} = -\sin\alpha\cos\varphi\cos\theta - \cos\alpha\sin\varphi, \tag{A8}$$

$$a_{22} = -\sin\alpha\cos\varphi\cos\theta + \cos\alpha\cos\varphi, \tag{A9}$$

$$a_{23} = \sin\alpha\sin\theta, \tag{A10}$$

$$a_{31} = \cos\varphi\sin\theta, \tag{A11}$$

$$a_{32} = \sin\varphi\sin\theta, \tag{A12}$$

$$a_{33} = \cos\theta, \tag{A13}$$

a_{IJ}^{-1} is the inverse matrices of a_{IJ} and given by

$$a_{IJ}^{-1} = \begin{pmatrix} a_{11} & a_{21} & a_{31} \\ a_{12} & a_{22} & a_{32} \\ a_{13} & a_{23} & a_{33} \end{pmatrix}. \tag{A14}$$

By using a_{IJ} and a_{IJ}^{-1} Eqs. (7)–(9) was derived.

REFERENCES

1. Williams, D. J., Nonlinear Optical Properties of Organic and Polymeric Materials, *ACS Symp. Ser.* 233, American Chemical Society, Washington D. C. (1983).
2. Chemla, D. S. and Zyss, J., *Nonlinear Optical Properties of Organic Molecules and Crystals*, Academic Press, Inc., Orland (1987).
3. Prasad, P. N. and Williams, D. J., *Introduction to Nonlinear Optical Effects in Molecules and Polymers*, John Wiley & Sons, Inc., New York (1991).
4. Watanabe, T., Yamamoto, H., Hosomi, T. and Miyata, S., New molecular design for noncentrosymmetric crystal structures: Lambda shape molecules for frequency doubling, in *Organic Molecules for Nonlinear Optics and Photonics* (eds. Messier, J., Kajzar, F. and Prasad, P.) 151, **E194**, Kluwer Academic Publishers, Dordrecht (1991).
5. Watanabe, T., Yoshinaga, K., Fichou, D., Chatani, Y. and Miyata, S., Strong second harmonic generation in poly(oxyethylene)/p-nitroaniline complexes crystallized under electric field, *Mat. Res. Soc. Symp. Proc.*, **109**, 339 (1988).

6. Watanabe, T., Yoshinaga, K., Fichou, D. and Miyata, S., Large second harmonic generation in electrically ordered p-nitroaniline-poly(oxyethylene) 'guest-host' systems, *J. Chem. Soc. Chem. Commun.*, 250 (1988).

7. Watanabe, T., Miyazaki, T. and Miyata, S., Nonlinear optics in host guest systems-crystalline polymer-dye binary mixtures, *MRS Int'l. Mtg. on Adv. Mats.*, 23 (1989).

8. Watanabe, T. and Miyata, S., Effect of crystallization process on the second harmonic generation of poly(oxyethylene)/p-nitroaniline systems, *SPIE Proc.*, **1147**, 101 (1989).

9. Meredith, G. R., VanDusen, J. G. and Williams, D. J., Optical and nonlinear optical characterization of molecularly doped thermotropic liquid crystalline polymers, *Macromolecules*, **15**, 1385 (1982).

10. Ye, C., Marks, T. J., Yang, J. and Wong, G. K., Synthesis of molecular arrays with nonlinear optical properties. Second-harmonic generation by covalently functionalized glassy polymers, *Macromolecules*, **20**, 2322 (1987).

11. Xu, C., Wu, B., Dalton, L. R., Ramon, P. M., Shi, Y. and Steier, W. H., New random main-chain, second-order nonlinear optical polymers, *Macromolecules*, **25**, 6716 (1992).

12. Watanabe, T., Zou, D., Shimoda, S., Tao, X., Usui, H., Miyata, S., Claude, C. and Okamoto, Y., A novel phase-matchig technique for a poled polymer waveguide, *Mol. Cryst. Liq. Cryst.*, **255**, 95 (1994).

13. Thackara, J. I., Lipscomb, G. F., Stiller, M. A., Ticknor, A. J. and Lytel, R., Poled electro-optic waveguide formation in thin-film organic media, *Appl. Phys. Lett.*, **52**, 1031 (1988).

14. Sugihara, H., Kinoshita, T., Okabe, M., Kunioka, S., Nonaka, Y. and Sasaki, K., Phase-matched second harmonic generation in poled dye/polymer waveguide, *Appl. Opt.*, **30**, 2957 (1991).

15. Singer, K. D., Sohn, J. E. and Lalama, S. J., Second harmonic generation poled polymer films, *Appl. Phys. Lett.*, **49**, 248 (1986).

16. Rikken, G., Seppen, C., Nijhuis, S. and Meijer, E., Poled polymers for frequency doubling for diode lasers, *Appl. Phys. Lett.*, **58**, 435 (1991).

17. Nalwa, H. S., Watanabe, T., Kakuta, A., Mukoh, A. and Miyata, S., N-phenylated aromatic polyurea: a new non-linear optical material exhibiting large second harmonic generation and U.V. transparency, *Polymer*, **34**, 657 (1993).

18. Mohlmann, G. R., Horsthuis, W. H .G., Mertens, J. W., Diemeer, M. B. J., Suyten, F. M. M., Hendriksen, B., Duchet, C., Fabre, P., Brot, C., Copeland, J. M., Mellor, J. R., Tomme, E. V., Daele, P. V. and Baets, R., Optically nonlinear polymeric devices, *SPIE Proc.*, **1560**, 426 (1991).

19. Lindsay, G. A., Henry, R. A., Hoover, J. M., Knoesen, A. and Mortazavi, M. A., Sub-Tg relaxation behavior of corona-poled optical polymer films and views on physical aging, *Macromolecules*, **25**, 4888 (1992).

20. Hayashi, A., Goto, Y., Nakayama, M., Sato, H., Watanabe, T. and Miyata, S., Novel photoactivatable nonlinear optical polymers: Poly[(((4-azidophenyl) carboxy) ethyl methacrylate], *Macromolecules*, **25**, 5094 (1992).

21. Dalton, L. R., Yu, L. P., Chen, M., Sapochak, L. S. and Xu, C., Recent advances and characterization of nonlinear optical materials second-order materials, *Synth. Metals*, **54**, 155 (1993).

22. Oudar, J. L. and Person, H. L., Second-order polarizabilities of some aromatic molecules, *Opt. Commun.*, **15**, 155 (1993).

23. Yamamoto, H., Katogi, S., Watanabe, T., Sato, H., Miyata, S. and Hosomi, T., New Molecular design approach for noncentrosymmetric crystal structures: Lambda (L)-shaped molecules for frequency doubling, *Appl. Phys. Lett.*, **60**, 935 (1992).

24. Nalwa, H. S., Watanabe, T. and Miyata, S., A comparative study of 4-nitroaniline, 1,5-diamino-2,4-dinitrobenzene and 1,3,5-triamino-2,4,6-trinitrobenzene and their molecular engineering for second-order nonlinear optics, *Opt. Mater.*, **2**, 73 (1993).

25. Zyss, J., Molecular engineering implications of rotational invariance in quadratic nonlinear optics: From dipolar to octopolar molecules and materials, *J. Chem. Phys.*, **98**, 6583 (1993).

26. Ulrich, R. and Torge, R., Measurement of thin film parameters with a prism coupler, *Appl. Opt.*, **12**, 2901 (1973).

27. Mortazavi, M. A., Knoesen, A., Kowel, S. T., Higgins, B. G. and Dienes, A., Second-harmonic generation and absorption studies of polymer-dye films oriented by corona-onset poling at elevated temperatures, *J. Opt. Soc. Am. B*, **6**, 733 (1989).

28. Kuzyk, M. G., Singer, K. D., Zahn, H. E. and King, L. A., Second-order nonlinear-optical tensor properties of poled films under stress, *J. Opt. Soc. Am. B*, **6**, 742 (1989).

29. Tao, X. T., Watanabe, T., Shimoda, S., Zou, D. C., Sato, H. and Miyata, S., L type main chain polymer for second harmonic generation, *Chem. Mater.*, **6**, 1961 (1994).

30. Tao, X. T., Watanabe, T., Zou, D. C., Shimoda, S., Sato, H. and Miyata, S., Polyurea with large positive birefringence for second harmonic generation, *Macromolecules*, **28**, 2637 (1995).

31. Jerphagnon, J. and Kurtz, S. K., Maker fringes: A detailed comparison of theory and Experiment for isotropic and uniaxial crystals, *J. Appl. Phys.*, **41**, 1667 (1970).

32. Tao, X. T., Watanabe, T., Zou, D. C., Ukuda, H. and Miyata, S., Phase-matched second-harmonic generation in poled polymers using positive birefringence, *J. Opt. Soc. Am. B*, **9**, 1581 (1995).

6. MOLECULAR DESIGN FOR STABILIZATION OF NONLINEAR OPTICAL DIPOLE MOMENTS IN POLYMERIC MATERIALS

NAOTO TSUTSUMI, OSAMU MATSUMOTO, and WATARU SAKAI

Department of Polymer Science and Engineering, Kyoto Institute of Technology, Matsugasaki, Sakyoku, Kyoto 606, Japan

1 INTRODUCTION

One of the features of organic and polymeric nonlinear optical (NLO) materials is the large electro-optic coefficients due to large second-order nonlinearity. Polymeric NLO materials have significant potential for the optical devices application based on the electro-optic modulation. Macroscopic noncentrosymmetry is required for the second harmonic generation (SHG) from these polymeric NLO materials. For these materials, electrical poling is a common procedure to break a center of symmetry of randomly oriented NLO dye molecules to achieve noncentrosymmetric alignment. The preferential alignment of NLO molecules was perturbed or disoriented by a molecular relaxation of materials, however, even if the sample was kept at the temperature below the glass transition temperature (T_g). Many efforts have been made to suppress the relaxation of NLO molecules in the polymeric matrix, for the purpose of fabricating thermally stable NLO polymeric materials. One approach is the use of crosslinking to suppress the segmental molecular motion of polymer matrix [1–6]. Another approach is the utilization of high glass transition temperature (T_g) material, such as polyimide [7].

Recently, we have synthesized a new type of NLO chromophore whose dipole moment is aligned transverse to the main chain backbone, and found that the resultant poled polymer film has large second harmonic efficiency with good thermal stability at the ambient condition. Present NLO chromophore is based on azobenzene dye. As pointed out by the previous work [8], NLO chromophore in this arrangement can be easier to orient by an external electric field than in structures where their dipole moments are pointing along the polymer backbone. Namely, the system whose NLO dipole moments aligned transverse to the polymer backbone requires the less deformation of the main chain backbone on orienting the dipole moment to the poling field direction than in the system whose NLO dipole moments is incorporated along the polymer backbone. In this article, we present the synthesis and the SHG properties of this novel type of NLO polymer material whose dipole moments are transversely aligned to the main chain backbone, and compare these properties with those for the polymer with NLO chromophore in the main chain backbone.

2 EXPERIMENTAL

2.1 Sample Preparation

4-(2-Hydroxyethylamino)-2-(hydroxymethyl)-4′-nitroazobenzene (T-AZODIOL) was used as the NLO chromophore whose dipole moment is aligned transverse to the main chain backbone. T-AZODIOL was synthesized via 2 step reactions; First, 3-(2-hydroxylamino)-benzylalcohol (DIOL) was prepared from *m*-aminobenzyl alcohol with 2-chloroethanol. Then the coupling reaction of DIOL with diazotized *p*-nitroaniline gave rise to T-AZODIOL. 4-[N-(2-Hydroxyethyl)-N-methylamino]-3′-(hydroxymethyl)-azobenzene (AZODIOL) dye was the NLO chromophore monomer for preparing the polymer with NLO chromophore in the main chain backbone. AZODIOL was synthesized by the coupling reaction of *m*-aminobenzyl alcohol with diazotized N-(2-hydroxylethyl)-N-methylaniline. The details of preparing T-AZODIOL [9] and AZODIOL [6] will be published elsewhere.

T-polymer was prepared from T-AZODIOL with tolylene 2,4-di-isocyanate in dimethylacetamide at the temperature of 85°C for 20 min in nitrogen atmosphere. L-polymer was prepared from AZODIOL with tolylene 2,4-di-isocyanate using the same condition as T-polymer was prepared. Chemical structures of T-polymer and L-polymer are shown in Figure 1. Figure 2 illustrates the schematic pictures of aligned NLO chromophores induced by poling in T- and L-polymers. Schematic pictures also give the idea that the less deformation of the main chain in T-polymer than in L-polymer is expected.

Spin-casting technique was employed to process thin films for SHG measurements. Spun-cast films were corona-poled at an elevated temperature to orient NLO chromophore to the poling direction. The distance between the sample and 0.1 ϕ tungsten wire for corona poling was kept at 1.4 cm.

2.2 Polymer Characterization

The Maker fringe method [10,11] is employed to measure SHG intensity of the poled spun-cast films. Laser source is a continuum model Surelite-10 Q-switched Nd:YAG pulse laser with 1064 nm *p*-polarized fundamental beam (320 mJ maximum energy, 7 ns pulse width and 10 Hz repeating rate). The generated second harmonic (SH) wave is detected by a Hamamatsu model R928 photomultiplier. The SH signal averaged on a Stanford Research Systems (SRS) model SR-250 gated integrator and boxcar averager module is transferred to a microcomputer through a SRS model SR-245 computer interface module. The details of the experimental procedure are described in Refs. [12,13].

Ultraviolet-visible spectra of the films were measured on a Shimadzu model UV-2101PC spectrophotometer. *m*-Line method using prism coupling apparatus was employed to measure the refractive indices of materials. Laser sources are a polarized He–Ne laser (632.8 nm) and a laser diode (830 nm). The prism of TaFD21 (HOYA Glass) with high refractive index (1.92588 at 632.8 nm) and a spin-coated or cast film was coupled with air-gap. Guided-wave spectra (*m*-lines) were obtained to

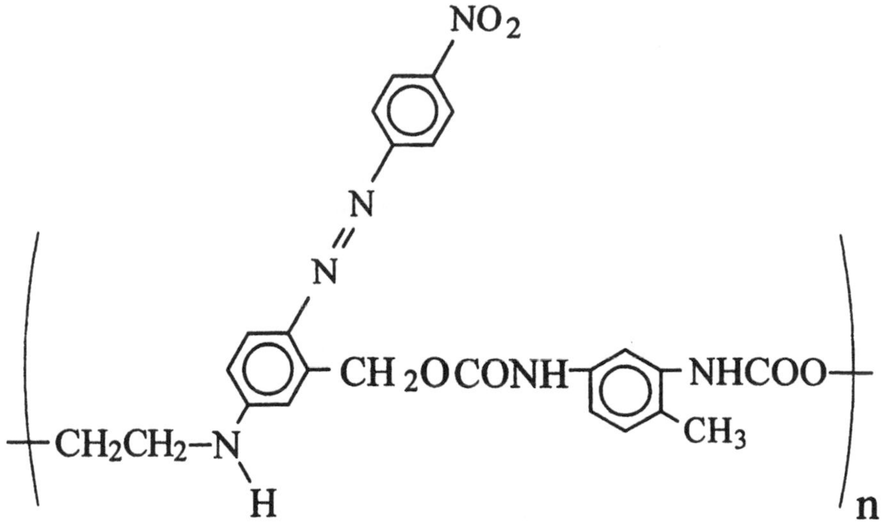

T-polymer

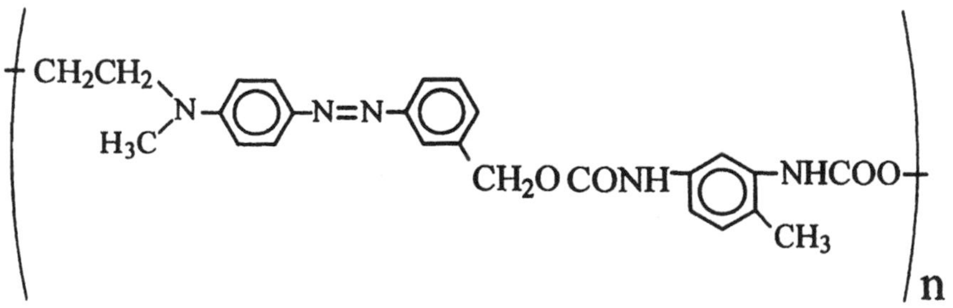

L-polymer

Figure 1 Chemical structures and codes of polymers.

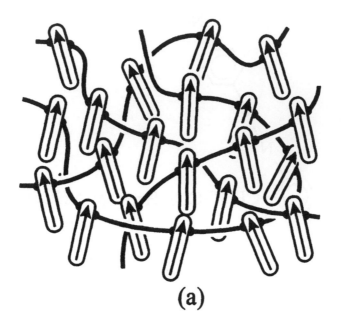

(a)

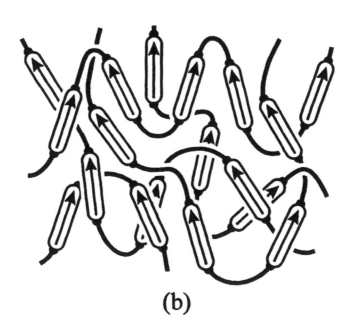

(b)

Figure 2 Schematic pictures of the alignment of NLO dipole moments in the polymer matrix. (a) in T-polymer and (b) in L-polymer.

determine the refractive indices. Differential scanning calorimetry (DSC) was carried out at a heating rate of 20°C/min in nitrogen atmosphere, using a Perkin Elmer DSC7 controlled by 1020 TA workstation. Density of the polymer film was measured in potassium iodide solution at 30°C using a sink and float test.

3 RESULTS AND DISCUSSIONS

3.1 Linear Optical Properties

Poling caused a decrease of absorption intensity and a shorter wavelength shift of absorption peak. Thermal annealing at the condition of the same temperature and time does not cause the change of intensity and spectral shift. Thus, the intensity change is ascribed to the orientation of the azobenzene dye to the direction of the film thickness induced by poling. The spectral blue shift has been reported for other cross-linked main-chain polymers [3], which is contrast to the red shift observed in most side-chain polymers [14].

The refractive indices (RI) for transverse electric field (TE) mode are measured using the m-line method at wavelength of 632.8 and 830 nm. RI values at wavelength of 632.8 and 830 nm were listed for the unpoled polymer films in Table 1. Wavelength dispersion of RI, $n_f(\lambda)$, can be fitted to a one-oscillator Sellmeier-dispersion formula,

$$n_f^2(\lambda) - 1 = \frac{q}{1/\lambda_0^2 - 1/\lambda^2} + A, \qquad (1)$$

where λ_0 is the absorption wavelength of the dominant oscillator, q is a measure for the oscillator strength, and A is a constant containing the sum of all the other oscillators. Figure 3 shows the plot of RI at the wavelength of 632.8 and 830 nm and the predicted curve of the wavelength dispersion of RI using Eq. (1) with $\lambda_0 = 470$ nm for T-polymer in (a) and $\lambda_0 = 414$ nm for L-polymer in (b). RI values at 532 and 1064 nm obtained from the predicted plots are listed in Table 1 for both T-polymer and L-polymer, which are used for the calculation of SHG coefficient.

Table 1 RI Values at wavelength of 632.8 and 830 nm measured and those at 532 and 1064 nm predicted by Eq. (1).

Polymer	Wavelength (nm)			
	632.8	830	532	1064
T-Polymer	1.6957	1.6686	1.779	1.661
L-Polymer	1.6828	1.6499	1.745	1.638

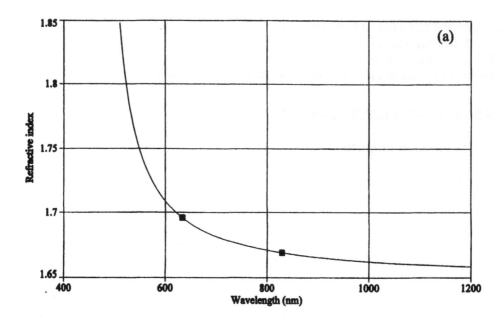

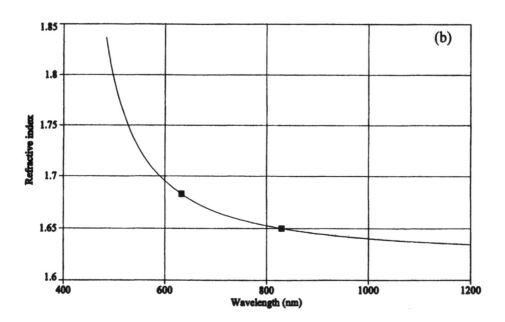

Figure 3   Plots of RI at the wavelength of 632.8 and 830 nm and the predicted plot curve of the wavelength dispersion of RI using Eq. (1) for T-polymer in (a) and for L-polymer in (b).

3.2   Determination of SHG Coefficients and Optimum Poling Condition for SHG

The SHG coefficients of the polymers are made relative to a Y-cut quartz plate ($d_{11} = 1.2 \times 10^{-9}$ esu (0.5 pm/ V)). The typical Maker fringe pattern could be observed for both the case of a p-polarized and a s-polarized fundamental beams.

It is important to optimize the poling condition (poling voltage, temperature and time) to obtain better SHG activities. For T-polymer, SHG coefficient increased with increasing applied voltage and leveled out at the voltage above 8.0 kV as shown in Figure 4(a). Increase of poling temperature causes the increase of SHG coefficient with large increment of coefficient at the temperature between 90°C and 95°C as shown in Figure 4(b). Poling time increased SHG coefficient with leveled out SHG coefficient at the time above 1 h as shown in Figure 4(c). When the sample film was poled at the optimum condition of poling voltage of 8 kV, temperature of 95°C and time of 60 min, d_{33} value of 1.6×10^{-7} esu (67 pm/ V) was obtained. This value is larger than SHG coefficient of lithium niobate ($LiNbO_3$).

For L-polymer, SHG coefficient increased with increasing applied voltage up to 7 kV with its decrease above 8 kV as shown in Figure 5(a). When poling temperature raises SHG coefficient increases as shown in Figure 5(b), but poling at the temperature above 70°C caused the film surface to be opaque.

In the theoretical expression of SHG coefficient d_{33} can be written as

$$d_{33} = \frac{N_d f_\omega^2 f_{2\omega} \beta \mu_g E_p}{10kT},$$

(2)

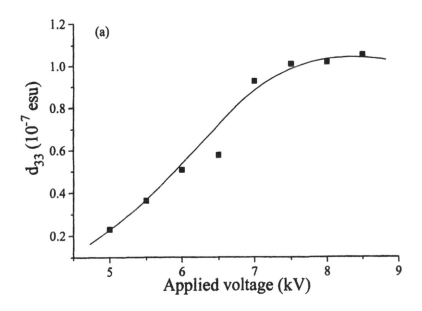

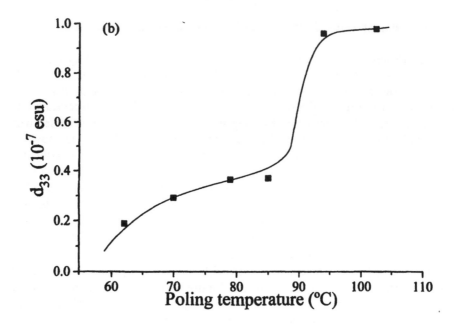

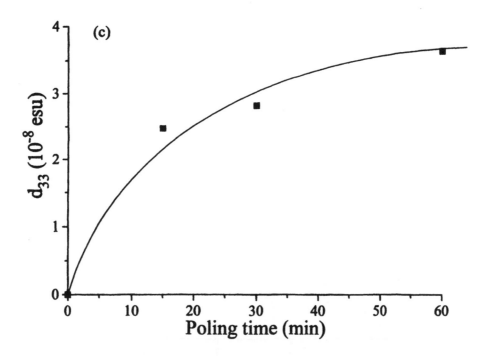

Figure 4 Effect of poling voltage in (a), temperature in (b) and time in (c) on SHG coefficient d_{33} for T-polymer.

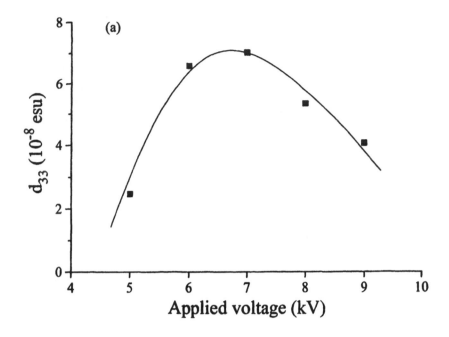

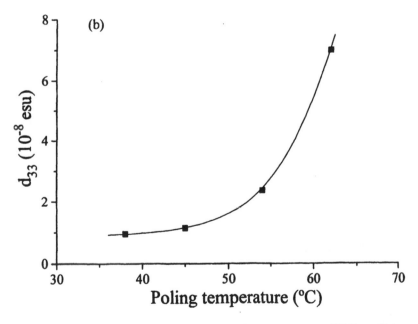

Figure 5 Effect of poling voltage in (a) and temperature in (b) on SHG coefficient d_{33} for L-polymer.

where N_d is the number density of noncentrosymmetric NLO molecules, β is the hyperpolarizability of the NLO guest, f_ω and $f_{2\omega}$ are Lorentz–Lorenz local field factors of the form $(\varepsilon + 2)/3$, μ_g is dipole moment of NLO chromophore at the ground state, E_p is poling electric field, k is Boltzmann's constant and T is the poling temperature. The value of ε has been taken as the square of the refractive index of the sample at either the fundamental or second harmonic frequency. The number of density N_d is calculated using the film density. To evaluate d_{33} values from Eq. (2), β, μ_g and E_p values must be estimated from the absorption spectra data.

E_p was determined from the intensity change in absorbance caused by the orientation of NLO chromophore, using the electrochromic theory [15,16]. The orientation-induced intensity change in absorbance can be related to the electric field, using the electrochromic theory,

$$\frac{A(p)}{A(0)} = 1 - G(u), \tag{3}$$

where $A(p)$ and $A(0)$ are the absorbance with and without electric field, respectively, and

$$G(u) = 1 - \frac{3\coth(u)}{u} + \frac{3}{u^2}, \tag{4}$$

$$u = \frac{\mu_g E_p}{kT}. \tag{5}$$

β can be calculated using the two level model [17,18],

$$\beta = \frac{9e^2}{4m}\left(\frac{h}{2\pi}\right)^2 \frac{W}{[W^2 - (2hv)^2][W^2 - (hv)^2]} f\Delta\mu, \tag{6}$$

where e is an elementary electric charge in esu, m is a rest mass of electron, h is Planck's constant, W is the energy at the absorption wavelength of the dominant oscillator, and hv and $2hv$ are the energies of the fundamental and second harmonic light. f is the oscillator strength of the dominant oscillator and can be evaluated from the absorption spectrum of the dominant oscillator [19,20]:

$$f = \frac{2303\,mc^2}{\pi e^2 Nn}\int \epsilon_{\tilde{v}}\,d\tilde{v} = 4.38 \times 10^{-9}\int \epsilon_{\tilde{v}}\,d\tilde{v}, \tag{7}$$

where c is the speed of light, N is Avogadro's number, n is the refractive index which is commonly omitted from the above expression, $\epsilon_{\tilde{v}}$ is molar extinction coefficient and the integration carried out over the absorption band of dominant oscillator.

β value is calculated using Eq. (6) with the dipole moment difference between ground and excited states $\Delta\mu\,(= \mu_e - \mu_g)$ estimated from the well-known azobenzene dyes [15]. E_p is evaluated from the decrease of the absorption maximum using Eq. (3). Then, for T-polymer, d_{33} can be calculated as 1.0×10^{-7} esu using E_p of

2.7 MV/cm from Eq. (3), μ_g of 8D [15] and β of 119×10^{-30} esu from Eq. (6), which can be comparable well to the experimentally obtained d_{33} was 0.91×10^{-7} esu. Experimentally obtained d_{33} value is in good agreement with theoretically calculated one.

3.3   Glass Transition Temperature for SHG Activity

Figure 6 shows the SHG activity profile *vs.* annealing temperature for T-polymer in (a) and L-polymer in (b), respectively. Successive SHG measurement at higher fixed temperature shown in the horizontal axis in the figure were carried out after both polymers were corona-poled. T-polymer does not show significant loss of SHG activity at the temperature up to 65°C and the large activity loss occurs at the temperature above 80°C; whereas, for L-polymer, the initial SHG loss starts at the temperature around 40°C. Significant large activity loss occurs at the temperature around 50–55°C. It is noted that these SHG activity profiles *vs.* annealing temperature can be compared to the SHG activity profiles when poling temperature is raised, which are shown in dotted lines in both Figures 6(a) and (b). Values of d_{33} on poling drastically increase when poling temperature increases from 85°C to 95°C for T-polymer and from 50°C to 60°C for L-polymer. Either poling temperature range in which the large increase of SHG activity occurs corresponds well to that the SHG activity is significantly lost on annealing for each polymer. These coincidences imply that both processes of the alignment (orientation) and randomization (reorientation) of NLO chromophore should be subjected to the same thermal activation. In other words, the thermal mobility for aligning NLO dipole moments on poling is the same as that for randomizing them on annealing. In this meaning, the glass transition temperatures for the SHG activities are involved in the temperature range between

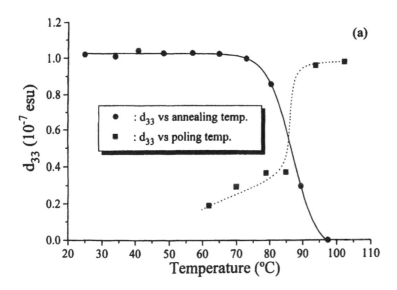

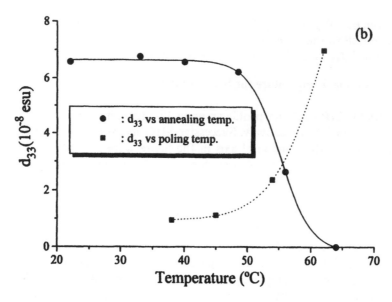

Figure 6 Temperature profile of d_{33} for T-polymer in (a) and for L-polymer in (b). SHG measurement was carried out at the fixed temperature shown on the horizontal axis in the figure. Dotted-curve is d_{33} profile on poling and the temperature on the horizontal axis in this curve is poling temperature.

80°C and 85°C for T-polymer and that between 50°C and 55°C for L-polymer. DSC measurement shows that T-polymer has T_g of 57°C and L-polymer has 90.7°C. These T_g values are not consistent with those estimated from the temporal thermal stability of SHG activity for both polymers shown above. T_g values measured by both DSC and SHG activities are compared in Table 2. These discrepancies may be related to the difference of mode in molecular motion which can be detected by each measurement. Furthermore, it is noted that T-polymer has better thermal stability around 60°C which is above T_g of 57°C, whereas L-polymer has less thermal stability at the temperature around 60°C in which the polymer is still in the glassy state. The details should be clarified by the further study using the dielectric relaxation and the thermal stimulated discharge current measurements.

Table 2 Comparison of T_g values measured by DSC and SHG activities.

	T_g (DSC) (°C)	T_g (SHG) (°C)
T-Polymer	57	80
L-Polymer	90.7	50

3.4   Thermal Stability of SHG Activity

Figure 7 shows the long-term thermal stability of d_{33} for both T-polymer and L-polymer when the sample films were stored at room temperature for the time which are shown in the horizontal axis in the figure. Plot in the figure is in the form of SHG coefficients d_{33} normalized by that measured at time of zero. It is noted that T-polymer has good long-term thermal stability of d_{33} (no significant relaxation at the ambient condition in 60 days) except for the small activity loss within a few days after poling, whereas the SHG coefficient d_{33} of L-polymer has largely decreased day by day and reached a half value of that at time of zero after two weeks storage. These storage time profile of SHG activity also supports that T-polymer has better thermal stability than L-polymer. Here one must remind that T_g values by DSC is 57°C for T-polymer and 90.7°C for L-polymer, therefore both polymers are in the glassy state at room temperature. Thus the difference of SHG activity profile against storage time between T- and L-polymers is due to the difference of the mobility of dipole moment between two polymers.

The question arising is what is the origin of stabilizing NLO dipole moments in T-polymer. One possibility is the free volume of matrix which provides the free space where the aligned dipole moment can be thermally reoriented. The density of T-polymer is larger than that of L-polymer as shown in Table 3. The larger density of T-polymer leads to the smaller free volume of T-polymer. The free volume V_f can

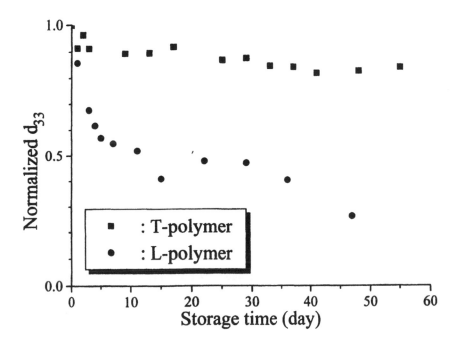

Figure 7   Long-term thermal stability of d_{33} values when T- and L-polymers are stored at room temperature.

be estimated from the experimentally obtained specific volume V_t (the reciprocal value of measured density) and the zero point molar volume V_0,

$$V_f = V_t - V_0, \tag{8}$$

where V_0 can be calculated from van der Waals volume V_w of polymer, $V_0 = 1.3V_w$ [21]. V_w of polymers was determined by the summation of the van der Waals volume of group contributions [22]. Table 3 shows the free volumes of T- and L-polymers calculated. As expected, V_f of T-polymer is smaller than that of L-polymer. Thus, the smaller free volume of T-polymer significantly contributes to the restriction of molecular motion in the glassy state of matrix. That is, the orientation of NLO chromophore in T-polymer is sustained by the smaller free volume of matrix in T-polymer.

3.5 Orientational Relaxation

The time-dependence of the decay profiles of the second-order nonlinear susceptibilities d_{33} for the corona poled thin films of both T- and L-polymers was investigated at various elevated temperatures. To monitor the decay of nonlinear susceptibilities d_{33} at given temperature, the film was heated as quickly as possible to the desired temperature and the start of the decay ($t = 0$ s) was taken as the time at which the film had reached and stabilized at the desired temperature. The analysis of the decay profiles provides the orientational relaxation of NLO chromophores at several elevated temperatures.

The observed time-dependence of decay of nonlinear susceptibilities d_{33} was nonexponential. The profiles of the time-dependent decay of NLO chromophores in polymers can in general be described by a Kohlrausch–Williams–Watts (KWW) stretched exponential form or by a sum of two exponentials fits, and both are often used to characterize the orientational relaxation of NLO chromophores in polymers. A sum of two exponentials fittings requires three parameters instead of two for KWW stretched exponential fittings. The time-dependent decay profiles for T- and L-polymers are well fitted by both fittings. Figure 8 shows the fitting examples for the relaxation of SHG activity of T-polymer at 25°C in (a) and at 80°C in (b). The physical model involving two distinct exponents of shorter and longer time constant leads to the complexities for characterizing the molecular motion of NLO chromophores in the amorphous polymer matrix. In the present case, a KWW stretched exponential fittings is employed to characterize the relaxation profiles of NLO

Table 3 Density and free volume (V_f) calculated by Eq. (8) for T- and L-polymers.

	Density (g/cm^3)	V_t (cm^3/g)	V_0 (cm^3/g)	V_f (cm^3/g)
T-polymer	1.246	0.803	0.632	0.171
L-polymer	1.147	0.872	0.662	0.210

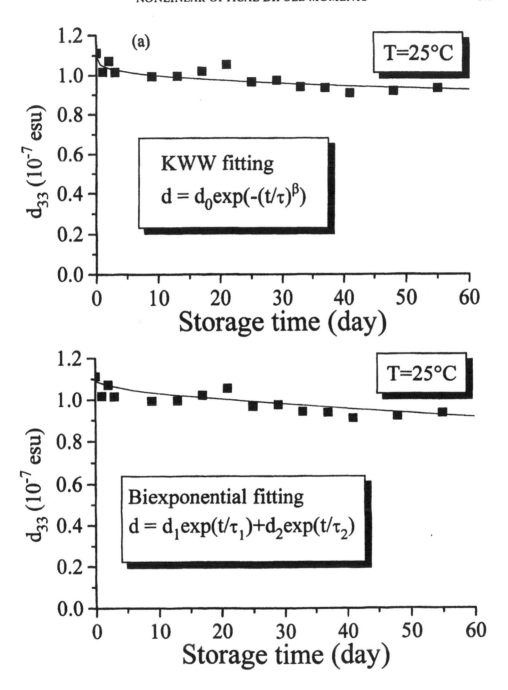

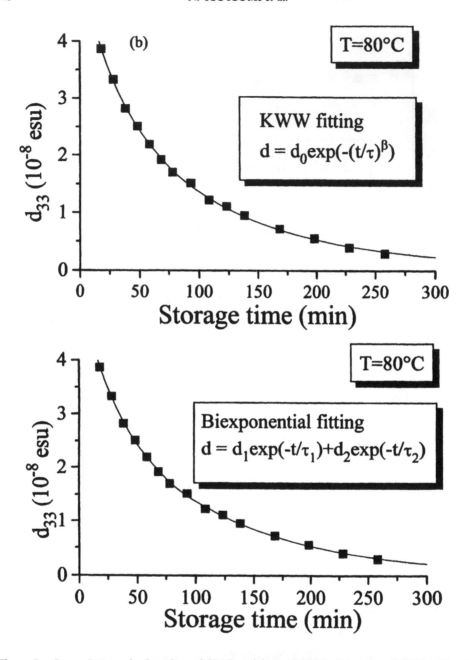

Figure 8 Curve fitting of relaxation of SHG activity at 25°C in (a) and at 80°C in (b) using KWW stretched exponential equation and biexponential equation. Parameters; (a) $d_0 = 1.10$, $\tau = 1.62 \times 10^9$ (s), $\beta = 0.31$; $d_1 = 0.04$, $d_2 = 1.06$, $\tau_1 = 4.05 \times 10^5$ (s) $\tau_2 = 3.79 \times 10^7$ (s) (b) $d_0 = 6.0$, $\tau = 3.40 \times 10^3$ (s), $\beta = 0.71$; $d_1 = 2.25$, $d_2 = 3.75$, $\tau_1 = 1.02 \times 10^3$ (s) $\tau_2 = 6.00 \times 10^3$ (s).

chromophores in T-polymer:

$$d = d_0 \exp(-(t/\tau)^\beta), \quad (0 < \beta \le 1), \qquad (9)$$

where τ is the characteristic relaxation time and β is a measure of width of the distribution of relaxation time and the extent of deviation from a single exponential behavior. When $\beta = 1$, the time-dependent decay profile corresponds to a single decay profile. The characteristic relaxation time τ is that time required for the system to decay to $1/e$ of its initial value. Figure 9 shows the KWW stretched exponential fittings of the time-dependence of decay profiles of d_{33} at several elevated temperatures for T-polymer. The vertical axis indicates the d_{33} values normalized by that at time $= 0$, $d_{33}(t)/d_{33}(0)$. β values for T-polymer are between 0.31 and 0.71, and increases with increasing temperature as shown in Figure 10. The relaxation time results exhibit that T-polymer has long-term thermal stability over 50 years at room temperature. The temperature dependence of the relaxation time is described well at all by a single energy-activated Arrhenius expression as shown in Figure 11. This result is in contrast to the fact that the guest–host system does not obey the single energy-activated Arrhenius plot [23,24].

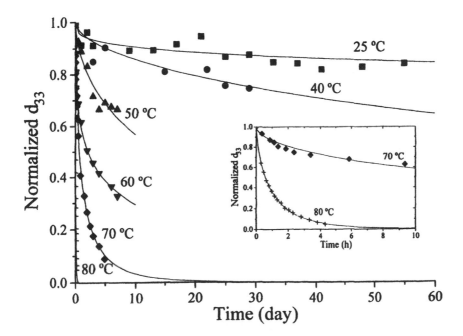

Figure 9 KWW stretched exponential fittings for the decay plot of SHG activity for T-polymer at different temperatures.

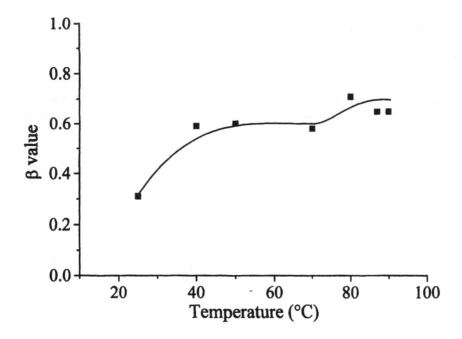

Figure 10   Plots of β value *vs.* temperature for T-polymer.

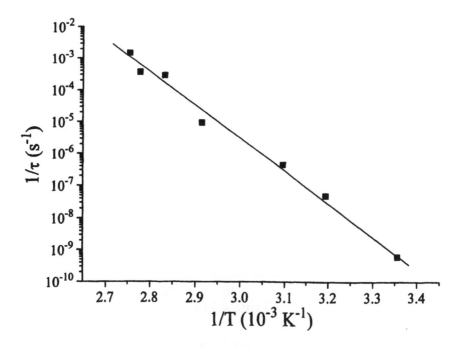

Figure 11   Arrhenius plot of SHG activity for T-polymer.

4 CONCLUSION

New approach of the molecular design for stabilizing NLO dipole moments in the polymeric materials are presented. The advantage of the utilization of the NLO unit whose one head is embedded in the polymer backbone and the dipole moments extends transverse to the main chain has been demonstrated. This polymer is amorphous with a high density of NLO chromophore moiety and optically transparent thin film can be processed by spin-casting. Poled T-polymer shows the large second order nonlinearity of $d_{33} = 1.6 \times 10^{-7}$ esu (67 pm/V). Good thermal stability of nonlinearity for T-polymer was observed at the ambient condition, which is contrast to the fact that L-polymer has the inferior thermal stability of nonlinearity even if it has higher T_g value than T-polymer. For both T- and L-polymers, the time dependence of the decay was nonexponential and well represented by the Kohlrausch–Williams–Watts stretched exponential function. The better thermal stability of T-polymer is related to the smaller free volume in T-polymer. This approach will open the way of fabricating electro-optic devices using this material with large second-order nonlinearity with good thermal stability. Next step is the creation of the uniform network polymers embedded the same type of NLO chromophore to achieve the thermally stable NLO active polymer system at higher elevated temperature.

REFERENCES

[1] D. Jungbauer, B. Reck, R. Twieg, D. Y. Yoon, C. G. Willson and J. D. Swalen, *Appl. Phys. Lett.* **56**, 2610 (1990).

[2] P. M. Ranon, Y. Shi, W. H. Steier, C. Xu, B. Wu and L. R. Dalton, *Appl. Phys. Lett.* **62**, 2605 (1993).

[3] J. A. F. Boogers, P. Th. A. Klaase, J. J. de Vliger and A. A. Tinnemans, *Macromolecules* **27**, 205 (1994).

[4] K. M. White, C. V. Francis and A. J. Isackson, *Macromolecules* **27**, 3619 (1994).

[5] N. Tsutsumi, S. Yoshizaki, W. Sakai and T. Kiyotsukuri, *Macromolecules* **28**, 6437 (1995).

[6] P. Prêtre, P. Kaats, A. Bohren, P. Günter, B. Zysset, M. Ahlheim, M. Stähelin and F. Lehr, *Macromolecules* **27**, 5476 (1994).

[7] R. D. Miller, D. M. Burland, M. Jurich, V. Y. Lee, C. R. Moylan, J. I. Thackara, R. J. Twieg, T. Verbiest and W. Volksen, *Macromolecules* **28**, 4970 (1995).

[8] C. Wender, P. Neuenschwander, U. W. Suter, P. Prêtre, P. Kaatz and P. Günter, *Macromolecules* **27**, 2181 (1994).

[9] N. Tsutsumi, O. Matsumoto, W. Sakai and T. Kiyotsukuri, *Macromolecules* **29**, 592 (1996); ibid **29**, 3338 (1996).

[10] P. D. Maker, R. W. Terhune, M. Nisenoff and C. M. Savage. *Phys. Rev. Lett.* **8**, 21 (1962).

[11] J. Jerphagnon and S. K. Kurtz, *J. Appl. Phys.* **40**, 1667 (1970).

[12] N. Tsutsumi, T. Ono and T. Kiyotsukuri, *Macromolecules* **26**, 5447 (1993).

[13] N. Tsutsumi, I. Fujii, Y. Ueda and T. Kiyotsukuri, *Macromolecules* **28**, 950 (1995).

[14] R. H. Page, M. C. Jurich, B. Reck, A. Sen, R. J. Twieg, J. D. Swalen, G. C. Bjorklund and C. G. Willson, *J. Opt. Soc. Am. B* **7**, 1239 (1990).

[15] W. Liptay, in *Excited States*, Vol. I, edited by E. C. Lim (Academic Press, New York and London, 1974) pp. 129–229.

[16] E. E. Havinga and P. van Pelt, *Ber. Bunsen-Ges. Phys. Chem.* **83**, 816 (1979).

[17] J. L. Oudar and D. S. Chemla, *J. Chem. Phys.* **66**, 2664 (1977).

[18] B. F. Levine and C. G. Bethea, *J. Chem. Phys.* **69**, 5240 (1978).

[19] N. J. Turro, in *Modern Molecular Photochemistry* (The Benjamin/Cummings Pub., Menlo Park, 1978) pp. 86–87.

[20] J. B. Birks, in *Photophysics of Aromatic Molecules* (John Wiley & Sons, London and New York, 1970) pp. 51–52.

[21] A. Bondi, in *Physical Properties of Molecular Crystals, Liquids and Glasses* (John Wiley & Sons, New York, 1968) chapters 3 and 4.

[22] D. W. van Krevelen, in *Properties of Polymers* (Elsevier, Amsterdam, 1990) chapter 4.

[23] M. Stähelin, D. M. Burland, M. Ebert, R. D. Miller, B. A. Smith, R. J. Twieg, W. Volksen and C. A. Walsh, *Appl. Phys. Lett.* **61**, 1626 (1992).

[24] A. Suzuki and Y. Matsuoka, *J. Appl. Phys.* **77**, 965 (1995).

7. POLYMER NETWORKS BASED ON ORGANOSOLUBLE POLYIMIDES EXHIBITING STABLE SECOND-ORDER NONLINEAR OPTICAL PROPERTIES

GING-HO HSIUE[a,*], JEN-KWAN KUO[a], RU-JONG JENG[b],
TE-HWEI SUEN[c], YING-LING LIU[a], and JOW-TSONG SHY[c]

[a] *Department of Chemical Engineering, National Tsing Hua University,
Hsinchu, Taiwan 300, R.O.C.;* [b] *Department of Chemical Engineering, National
Chung Hsing University, Taichung, Taiwan 400, R.O.C.;* [c] *Department of
Physics, National Tsing Hua University, Hsinchu, Taiwan 300, R.O.C.*

ABSTRACT

We describe here the preparation and characterization of a new class of nonlinear optical (NLO) crosslinked polymer network based on organosoluble polyimides. Sol–gel reaction of a nonlinear optically active alkoxysilane dye (ASD) is utilized to create a polymer network. Excellent long-term stability of the effective second harmonic coefficients (d_{eff}) at 120°C is observed because of the features of high glass transition temperature and extensive crosslinking. An interpenetrating polymer network (IPN) system consisting of a polyimide/ASD and a crosslinkable NLO-active polymer is also reported.

INTRODUCTION

Nonlinear optical (NLO) polymers have shown increased potential in practical applications, such as electro-optic (EO) modulation [1]. For integrated optic and EO applications, work needs to be done to improve thin film technology as it relates to orientation phenomena, film optical uniformity and quality, and, most importantly, long-term NLO stability up to 100°C and short excursion to the temperatures of 250°C or higher [2]. To preserve second-order NLO properties, NLO chromophores are usually incorporated into either high-T_g polymers and/or crosslinked polymer network [3–17].

Aromatic polyimides are well known for their high thermal stability and high glass transition temperatures (T_g) [18]. One approach of improving NLO stability at elevated temperature is to dissolve NLO chromophores in polyimide matrices [15]. The orientation of the chromophores is preserved due to the decreased mobility of the host polymer and the chromophores. More recently, Lin *et al.* first developed a chromophore-functionalized polyimide [19]. This approach of covalently bonding NLO chromophores to polyimide backbone provided a versatile synthesis in obtaining NLO-active polyimide based materials with large and stable second-order NLO properties [20–22]. Moreover, polyimide was used as an organic host and a polyimide/inorganic composite was prepared as a second-order NLO material via *in situ* polymerization of ASD [9]. ASD was added to polyamic acid derived from methylene dianiline and 3,3′,4,4′-benzophenonetetracarboxylic acid anhydride.

---

* *Corresponding author.*

Following thermal curing and poling, transparent polyimide films containing aligned NLO moieties in the inorganic network were obtained. This polyimide composite containing the ordered linked alkoxysilane dye can also be classified as a semi-interpenetrating polymer network (Semi-IPN-II) [17].

High glass transition temperature and extensively crosslinked network of the material are essential for the excellent long-term second-order NLO properties at elevated temperature. An NLO active interpenetrating polymer network (IPN) developed by Marturunkakul *et al.* was proven to meet the aforementioned criterion [23]. This IPN system happens to have the characteristics of the entanglements of the crosslinked network. The entanglement of two rigid networks are able to suppress the motion of each polymer effectively. By comprising two rigid components, the IPN can also possess an extremely high glass transition temperature.

Based on above information, hybridizing high T_g and crosslinked network characteristics seems to be a valid way for improving the temporal stability of NLO polymeric system. As mentioned earlier, polyimide is well known for its high temperature properties. However, most polyimides are not thermoplastics or organosoluble, and imidization of polyamic acid and their densification take place at relatively high temperature for an extended period of time [18]. These two characteristics are detrimental to the poling process. Yet, several researchers have synthesized organosoluble polyimides [24–27]. These organosoluble polyimides have been shown to possess T_g's above 400°C. Via organosoluble polyimides, improved processability can be achieved without sacrificing useful properties. Recently, Lakshmanan *et al.* have synthesized a series of aromatic hydroxyl-terminated organosoluble polyimides [28]. The molecular weights and T_g's were controlled by changing the feed ratio of an end-capper and the monomers. Moreover, the hydroxyl functionalities at the chain ends provides the possibility for further chemical incorporation.

According to literature [29], alkoxysilane can react with phenol groups to form phenoxy compound. In one example, the properties of an alkoxysilane covalently bonded to an NLO chromophore (ASD), which was reacted with 1,1,1-tris(4-hydroxyphenyl)ethane to form a crosslinked phenoxysilicon material, were investigated [14]. Upon heating to high temperatures, the rigid multiphenoxyl molecule containing three phenol groups and alkoxysilane with three methoxyl groups lead to a highly crosslinked network (Figure 1). In this paper, we adopted similar chemistry to prepare an NLO polymer network by taking advantage of the hydroxyl functionalities of the organosoluble polyimides. In some polyimides, the reactive groups are located not only at chain ends, but at main chains as well. More reactive groups at polyimide chains are preferred because it provides for more extensive chemical incorporation. An NLO active ASD, serving as a crosslinking agent, was added to the organosoluble polyimide. The ASD was then reacted with aromatic hydroxyl groups to form an NLO active crosslinked polyimide containing phenoxysilicon linkages (Scheme II). Furthermore, an IPN consisting of a polyimide/ASD and a NLO active crosslinkable polymer (Polymer 11) [30] was also investigated for its NLO properties.

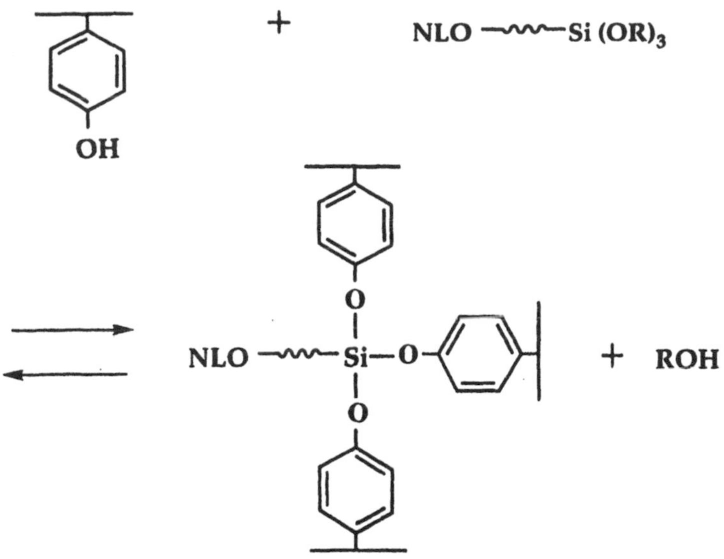

Figure 1 Formation of phenoxysilicon network.

EXPERIMENTAL

Aromatic organosoluble Polyimide A (PI-A), Polyimide B (PI-B), Polyimide C (PI-C), and Polyimide D (PI-D) (Figure 2) with various molecular weights were successfully synthesized by solution imidization techniques [28]. There is no hydroxyl group on PI-A, whereas the hydroxyl groups are located at the chain ends of PI-B. In addition to the hydroxyl groups located at chain ends, PI-C and PI-D also have lateral hydroxyl groups at backbones. Especially, PI-D possesses two hydroxyl groups on each repeat unit. α,α'-Bis-(4-aminophenol)-1,4-di-isopropylbenzene (Bis-P), 4,4'-oxydiphthalic anhydride (ODPA), 2,2-Bis(3-amino-4-hydroxyphenyl)-hexafluoropropane (BAPAF), p-aminophenol (P-AP), and aniline are available from TCI Co. and used as received. P-AP and aniline serve as the end capper agent. T_g's were determined using differential scanning calorimetry (DSC) with a heating rate of 10°C/min. NLO chromophore, 4-((4'-nitrophenyl)azo) aniline was coupled with (3-glycidoxypropyl) trimethoxysilane to form ASD (Figure 3(a)) [14]. The thermal degradation temperature (T_d) of materials were determined on a thermogravimetric analyzer (TGA) with a heating rate of 10°C/min under air. IPN samples were prepared with the weight ratios of 1:1, 1:3, 1:5 (Polymer 11 to PI-D/ASD), respectively. Chemical structure of Polymer 11 is shown in Figure 3(b). *In situ* poling technique was used to impart the second-order NLO properties to the polymer film. The applied current was maintained at 1μA with a potential of 4 kV while the poling temperature was kept at 240°C for 10 min. Network formation and alignment of chromophores were completed simultaneously during this period. Second-order NLO properties of samples were measured by second harmonic generation at 1.06 μm [6].

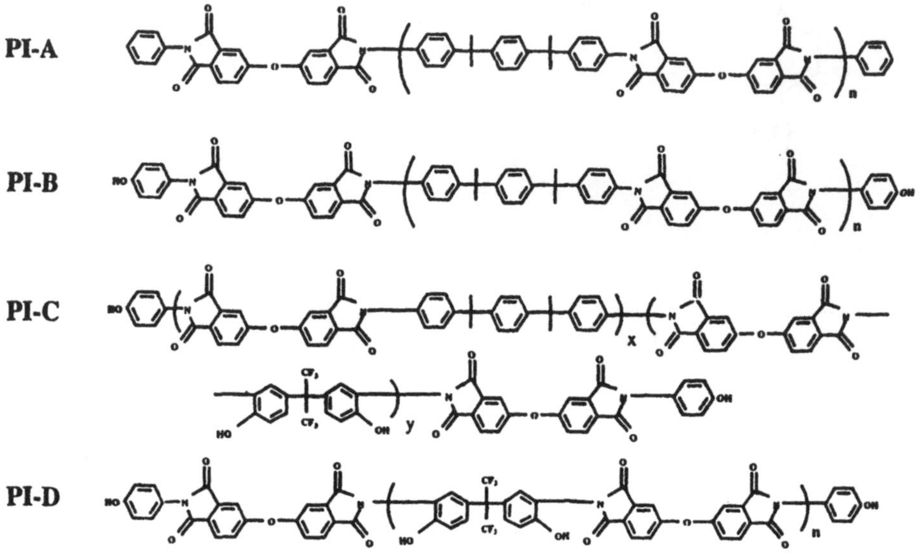

PI-A

PI-B

PI-C

PI-D

Figure 2   Chemical structures of PI-A, PI-B, PI-C and PI-D.

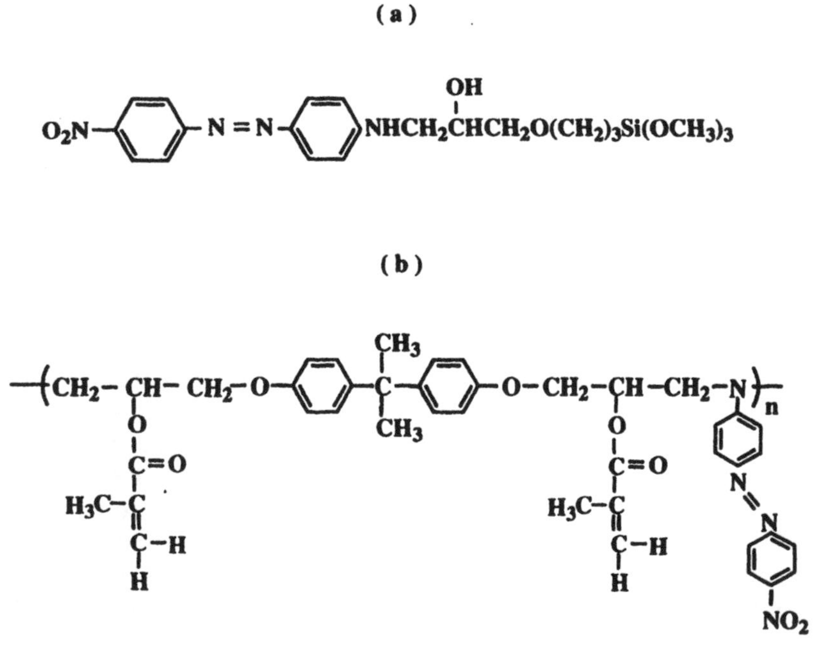

(a)

$$O_2N-\underset{}{\bigcirc}-N=N-\underset{}{\bigcirc}NHCH_2\overset{\overset{OH}{|}}{C}HCH_2O(CH_2)_3Si(OCH_3)_3$$

(b)

Figure 3   Chemical structures of: (a) alkoxysilane dye (ASD); and (b) Polymer 11.

RESULTS AND DISCUSSION

PI-A, PI-B, PI-C, and PI-D were soluble in various organic solvents such as chloroform, tetrahydrofuran, N-dimethylacetamide, acetone, etc. The T_g's of these polyimides were tailored in the range of 169–250°C as the molar feed ratio of the end-capper to the monomers changed from 0.5 to 0.05. The methoxyl groups of ASD was reacted with this aromatic hydroxyl groups of PI-B, PI-C, or PI-D to form crosslinked NLO-active polyimides (Figure 4) [31], whereas the sol–gel reaction of ASD occurred within PI-A matrices to form an organic/inorganic composite [9]. PI-A, PI-B, PI-C, and PI-D, with respective T_g's of 200, 195, 197, and 200°C, were selected for further study. Properties of these polyimides are shown in Table 1. The molecular weights of PI-C and PI-D are slightly higher than those of PI-A and PI-B in spite that the T_g's of all the polyimides are tailored to be about 200°C. The optimum curing condition of polyimide/ASD samples was chosen to be at 240°C for 10 min due to the consideration of the thermal stability of ASD ($T_d = 250$°C) and effective alignment of the NLO chromophores. Various wt% of ASD loading in polyimides were experimented to obtain the optimized wt% of ASD loading in each particular polyimide. T_g's for the cured polyimide/ASD samples with various wt% of ASD loadings are shown in Figure 5. The T_g of each polyimide increases as the loading of ASD increases. However, the loading levels are limited to 30–40%

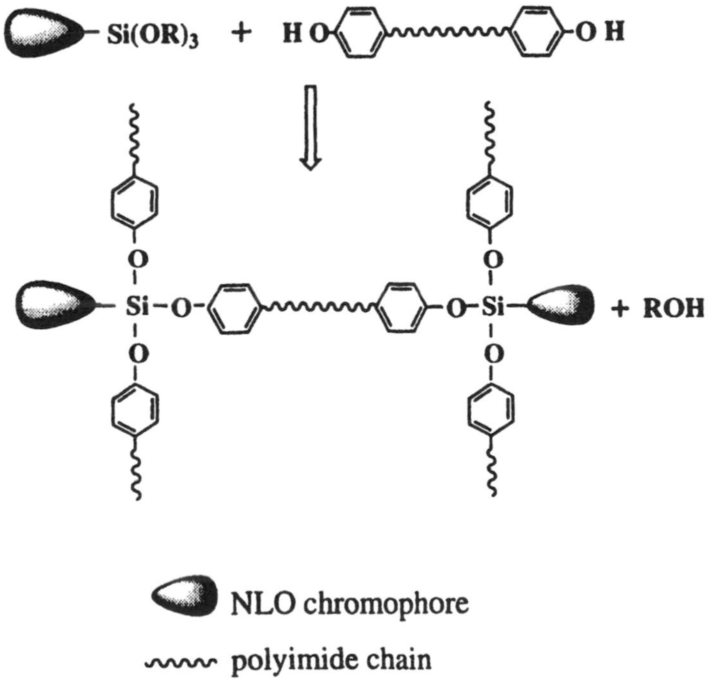

Figure 4 Formation of a crosslinked polyimide network.

Table 1   Properties of organosoluble polyimides.

Polymer	Properties			
	T_g (°C)	$\bar{M}_w$	T_d (°C)[a]	OH content[b]
PI-A	200	7980	453	0
PI-B	195	8050	450	0.61
PI-C	197	9400	440	1.98
PI-D	200	9000	431	3.69

[a] 5% weight loss.
[b] mmole/g; acetyl chloride method.

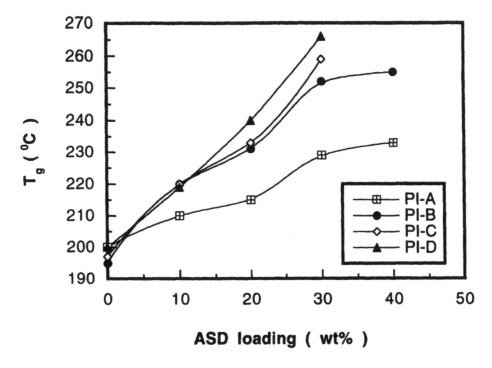

Figure 5 T_g's for the cured polyimides/ASD samples with various wt% of ASD loading.

because of the appearance of ASD aggregation. Based on the aforementioned study, the optimized wt% of ASD loading for all of the polyimide/ASD samples is chosen to be 30% (unless otherwise stated, this wt% of ASD loading was used throughout the study). DSC thermograms for pristine polyimide sample and cured polyimide/ASD samples are illustrated in Figure 6. The T_g of the cured polyimide/ASD increases as the content of hydroxyl groups for polyimides increases. In addition, ΔC_p (T_g) of the cured polyimide/ASD samples decreases as the T_g rises up. This

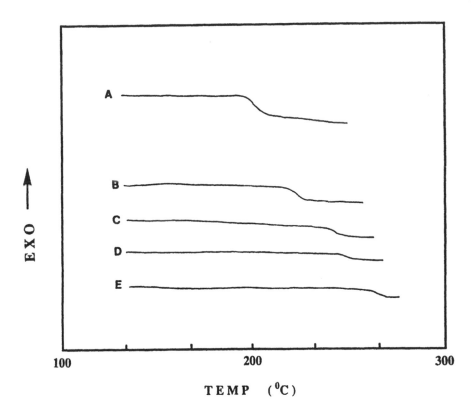

Figure 6 DSC thermograms of: (A) pristine PI-A, (B) PI-A/ASD, (C) PI-B/ASD (D) PI-C/ASD, (E) PI-D/ASD.

indicates a net decrease in both chain mobility and vibrational contributions to C_p as a result of the increased interchain crosslinking density [32,33]. It is important to note that all of the organosoluble polyimides remained soluble even after being subjected to 350°C for 5 min. Their T_g's remained unchanged after this thermal treatment. Therefore, the increase in T_g's of the cured Polyimide/ASD samples is solely due to the formation of the silicon containing network.

The formation of phenoxysilicon linkages was analyzed by infrared (IR) spectroscopy. Figure 7 shows the IR spectra of PI/ASD samples before and after curing. For the cured PI-A/ASD sample, the broadening absorption peak at around 1100 cm^{-1} was identified as the formation of Si—O—Si network resulted from sol–gel reaction among ASD molecules [9]. The appearance of phenoxysilicon bond absorption at 960 cm^{-1} for the cured PI-B/ASD indicates the inter-polymer chain crosslinking reaction have occurred. Moreover, the absorption peak of hydroxyl stretching around 3400 cm^{-1} decreased drastically after curing. This further supports the formation of phenoxysilicon linkages. On the other hand, the cured PI-B/ASD sample displays a narrower peak at around 1100 cm^{-1}. This means the formation of Si—O—Si network is less extensive in the cured PI-B/ASD system as compared to

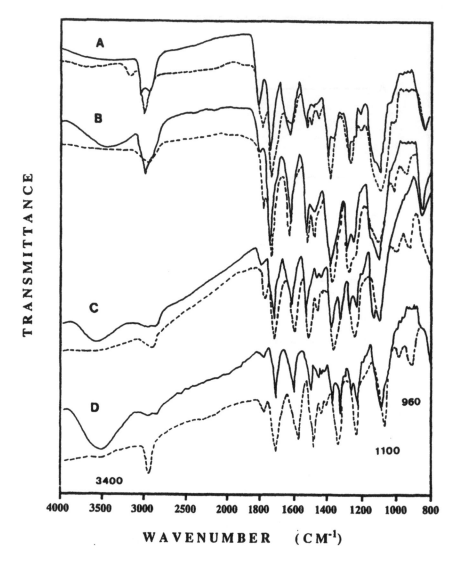

Figure 7 IR spectra of polyimides/ASD. (A) (PI-A/ASD), (B) (PI-B/ASD), (C) (PI-C/ASD), (D) (PI-D/ASD); from top to bottom: pristine, cured.

the cured PI-A/ASD sample. For the cured PI-C/ASD or PI-D/ASD sample, a stronger absorption peak at $960 \, cm^{-1}$ was observed. This indicates that more extensive network of phenoxysilicon linkages has formed due to the relative abundance of reactive groups.

To investigate the absorption behavior as a functional of time, the absorption spectrum (Figure 8) was taken regularly over a 168-hour period under thermal treatment at 120°C for the poled/cured PI-B/ASD sample. Immediately after poling

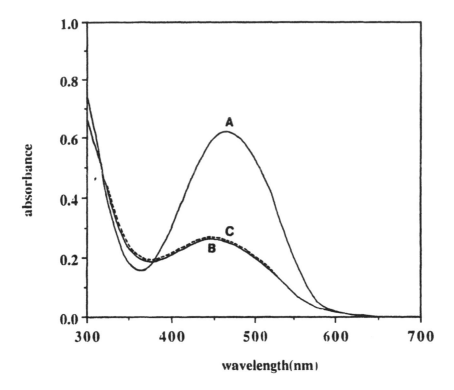

Figure 8 UV–Vis spectra of PI-B/ASD sample: (A) pristine sample; (B) immediately after poling and curing; (C) poled/cured sample, after thermal treatment at 120°C for 168 h.

and curing, a decrease in absorbance was observed. This is due to some dye degradation and orientational dichroism [34]. During the next 168 h the absorption spectra remained unchanged.

Properties of the poled and cured polyimide/ASD samples are shown in Table 2. The poled/cured film of PI-A/ASD has a d_{33} value of 38 pm/V. After thermal treatment at 120°C for 168 h, a reduction of 28% in d_{eff} value was observed for the poled/cured sample. On the other hand, the poled/cured film of PI-B/ASD has a d_{33} of 20 pm/V. After the aforementioned thermal treatment, 20% of decay in d_{eff} was observed. Compared with poled/cured PI-A/ASD ($T_g = 229$°C), poled/cured PI-B/ASD ($T_g = 240$°C) exhibits better temporal stability. This is a direct consequence of inter-polymer chains crosslinking in the cured PI-B/ASD.

To further enhance temporal stability, incorporation of reactive groups onto polyimide main chains seems to be a reasonable approach. Polyimides with multi-hydroxyl groups on the main chains were obtained using diamine monomer containing hydroxyl groups. After curing at 240°C for 10 min, T_g's of PI-C/ASD and PI-D/ASD advanced to 259 and 266°C, respectively. The polyimide with more reactive groups resulted in higher crosslinking density as evidenced by the higher T_g and smaller ΔC_p (T_g) of PI-D/ASD. The poled and cured films of PI-C/ASD and

Table 2   Properties of poled and cured polyimide/ASD samples.

Polymer	Properties			
	T_g (°C)	T_d (°C)	d_{33}(pm/V)	Relaxation[b]
PI-A/ASD[a]	229	480	38	0.72
PI-B/ASD	240	500	20	0.80
PI-C/ASD	259	503	17	0.92
PI-D/ASD	266	501	15	0.97

[a] Cured and poled samples with 30 % wt of ASD.
[b] $d_{eff}(t)/d_{eff}(0)$ values under 120°C after 168 h.

PI-D/ASD exhibit d_{33} values of 17 and 15 pm/V, respectively. Insignificant decay (less than 10%) in d_{eff} for PI-C/ASD and PI-D/ASD have been observed after thermal treatment at 120°C for 168 h. The temporal stability of the poled/cured PI-C/ASD and PI-D/ASD are better than that of the poled/cured PI-B/ASD. This indicates that the extensive inter-polymer chains crosslinking in a high T_g polymer would further upgrade temporal stability.

PI-D/ASD was chosen to be one of the IPN components due to its highest T_g and crosslinking density among the organosoluble polyimides in this work. The other IPN component is Polymer 11 whose poled/cured sample exhibits large optical nonlinearity (57 pm/V). NLO active IPN samples with different weight ratios of PI-D/ASD to Polymer 11 were obtained via simultaneous curing of these two components. The properties of poled/cured IPN samples are shown in Table 3. T_g's were not observed for all the IPN samples, whereas T_d increased with the increased content of PI-D/ASD in IPN samples. It is important to note that the homogeneity of this IPN is confirmed using SEM. No sign of any phase separation was observed when the magnification was increased upto 40 K. Optical microscopy reveals a clear transparent featureless film as well. The d_{33} for the IPN samples were found to be

Table 3   Properties of poled/cured IPN samples.

Polymer[a]	Properties		
	T_g (°C)	T_d (°C)	d_{33}(pm/V)
POLYMER 11	130	270	57
IPN-1[b]	—	350	34
IPN-2[c]	—	400	28
IPN-3[d]	—	425	20
PI-D/ASD	266	501	15

[a] Samples have been cured and poled.
[b] Polymer 11:PI-D is 1:1.
[c] Polymer 11:PI-D is 1:3.
[d] Polymer 11:PI-D is 1:5.

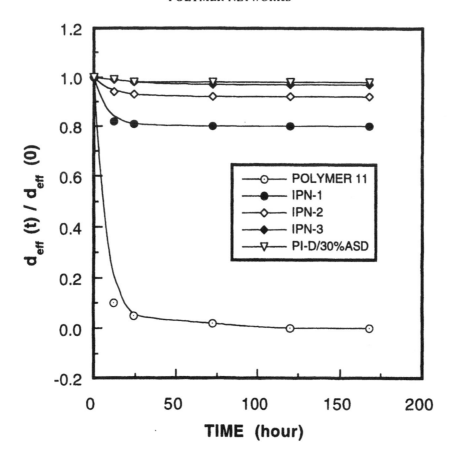

Figure 9 Time behavior of the effective second harmonic coefficients for IPN samples under 120°C thermal treatment.

ranged from 34 to 20 pm/V. These d_{33} are larger than those of the poled/cured polyimide/ASD samples, while the temporal stability maintains almost the same as that of the polyimide/ASD samples (Figure 9).

CONCLUSION

We have shown a novel class of NLO active crosslinked network based on organosoluble polyimides and ASD. Excellent long-term NLO stability at 120°C is a result of the features of high T_g and inter-polymer chain crosslinking. However, the stability of the second-order NLO properties is limited by the NLO chromophore degradation temperature. An alkoxysilane dye with a NLO chromophore capable of withstanding higher cure temperature provides the possibility for thin films with even more stable second-order nonlinearity.

ACKNOWLEDGMENT

We thank the National Science Council of Taiwan, ROC for financial support (Contract No. NSC-85-2216-E-007-046).

REFERENCES

1. S. Ermer, J. F. Valley, R. Lytel, G. F. Lipscomb, T. E. Van Eck and D. G. Girton, *Appl. Phys. Lett.*, **61** (1992) 2272 .
2. R. D. Miller, D. M. Burland, M. Jurich, V. Y. Lee, C. R. Moylan, J. I. Thackara, R. J. Twieg, T. Verbiest and W. Volksen, *Macromolecules*, **28** (1995) 4970.
3. C. Xu, B. Wu, O. Todorova, L. R. Dalton, Y. Shi, P. M. Ranon and W. H. Steier, *Macromolecules*, **26** (1993) 5303.
4. M. A. Hubbard, T. J. Marks, J. Yang and G. K. Wong, *Chem. Mater.*, **1** (1989) 167.
5. M.Eich, B. Reck, D. Y. Yoon, C. G. Willson and G. C. Bjorklund, *J. Appl. Phys.*, **66** (1989) 3241.
6. R. J. Jeng, Y. M. Chen, J. Kumar and S. K. Tripathy, *J. Macromol. Sci., Pure Appl. Chem.*, **A29** (1992) 1115.
7. R. J. Jeng, Y. M. Chen, A. K. Jain, S. K. Tripathy and J. Kumar, *Opt. Commun.*, **89** (1992) 212.
8. R. J. Jeng, Y. M. Chen, A. K. Jain, J. Kumar and S. K. Tripathy, *Chem. Mater.*, **4** (1992) 972.
9. R. J. Jeng, Y. M. Chen, A. K. Jain, J. Kumar and S. K. Tripathy, *Chem. Mater.*, **4** (1992) 1141.
10. M. Chen, L. R. Dalton, L. P. Yu, Y. Shi and W. H. Steier, *Macromolecules*, **25** (1992) 4032.
11. C. Xu, B. Wu, L. R. Dalton, Y. Shi, P. M. Ranon and W. H. Steier, *Macromolecules*, **25** (1992) 6714.
12. Y. Jin, S. H. Carr, T. J. Marks, W. Lin and G. K. Wong, *Chem. Mater.*, **4** (1992) 963.
13. S. Marturunkakul, J. I. Chen, L. Li, R. J. Jeng, J. Kumar and S. K. Tripathy, *Chem. Mater.*, **5** (1993) 592.
14. R. J. Jeng, Y. M. Chen, J. I. Chen, J. Kumar and S. K. Tripathy, *Macromolecules*, **26** (1993) 2530.
15. J. W. Wu, J. F. Valley, S. Ermer, E. S. Binkley, J. T. Kenney, G. F. Lipscomb and R. Lytel, *Appl. Phys. Lett.*, **58** (1991) 225.
16. J. A. F. Boogers, P. Th. A. Klaase, J. J. de Vlieger and A. H. A. Tinnemans, *Macromolecules*, **27** (1994) 205.
17. S. Marturunkakul, J. I. Chen, R. J. Jeng, S. Sengupta, J. Kumar and S. K. Tripathy, *Chem. Mater.*, **5** (1993) 743.
18. P. M. Cotts, W. Volken, in *Polymers in Electronics*, ed. by T. Davidson, ACS Symposium Series 242 (Washington D. C., 1984), 227.
19. J. T. Lin, M. A. Hubbard, T. J. Marks, W. Lin and G. K. Wong, *Chem. Mater.*, **4** (1992) 1148.
20. A. K. Y. Jen, Y. J. Liu, Y. Cai, V. P. Rao and L. R. Dalton, *J. Chem. Soc. Chem. Commun.* (1994) 2711.
21. D. Yu, A. Gharavi and L. Yu, *Macromolecules*, **28** (1995) 784.
22. T. Verbiest, D. M. Burland, M. C. Jurich, V. Y. Lee, R. D. Miller and W. Volksen, *Macromolecules*, **28** (1995) 3005.

23. S. Marturunkakul, J. I. Chen, L. Li, R. J. Jeng, J. Kumar and S. K. Tripathy, *Chem. Mater.*, **5** (1993) 592.
24. C. P. Yang and Y. Y. Yen, *J. Polym. Sci. Polym. Chem.*, **30** (1992) 1855.
25. P. P. Huo and P. Cebe, *Polymer*, **34** (1993) 696.
26. T. M. Moy, C. D. Deporter and J. E. McGrath, *Polymer*, **34** (1993) 819.
27. M. E. Rogers, M. H. Brink and J. E. McGrath, *Polymer*, **34** (1993) 849.
28. P. Lakshmanan, S. Srinivasan, T. Moy and J. E. McGrath, *Polym. Prepr. Am. Chem. Soc. Div. Polym. Chem.*, **34**(1) (1993) 707.
29. V. Bazant, V. Chralovsky and J. Rathousky, *Organosilicon Compounds* (Academic Press, New York, 1965), p. 58.
30. S. Marturunkakul, J. I. Chen, L. Li, X. L. Jiang, R. J. Jeng, S. K. Sengupta, J. Kumar and S. K. Tripathy, *Polym. Prepr. Am. Chem. Soc. Div. Polym. Chem.*, **35**(2) (1994) 134.
31. G. H. Hsiue, J. K. Kuo, R. J. Jeng, J. I. Chen, X. L. Jiang, S. Marturunkakul, J. Kumar and S. K. Tripathy, *Chem. Mater.*, **6** (1994) 884.
32. G. Stevens and M. Richardson, *Polymer*, **24** (1983) 851.
33. D. Plazek and Z. Frund, *J. Polym. Sci. Polym. Phys.*, **28** (1990) 431.
34. M. A. Mortazavi, A. Knoesen, S. T. Kowel, B. G. Higgins and A. Dienes, *J. Opt. Soc. Am. B*, **6** (1989) 773.

8. RELAXATION PROCESSES IN NONLINEAR OPTICAL SIDE-CHAIN POLYIMIDE POLYMERS

P. KAATZ, Ph. PRÊTRE, U. MEIER, U. STALDER, Ch. BOSSHARD*,
and P. GÜNTER

*Nonlinear Optics Laboratory, Institute of Quantum Electronics,
ETH Hönggerberg, CH-8093 Zürich, Switzerland*

ABSTRACT

Polyimide side-chain polymers have been prepared with high glass transition temperatures $T_g \approx 170°C$, high nonlinearities ($d_{33} = 69$ pm/V @ $\lambda = 1.3\,\mu m$), and electro-optic coefficients r_{33} up to 20 pm/V in the infrared. The relaxation mechanisms of the side-chain chromophores have been investigated below and above the glass transition by second-harmonic decay, dielectric relaxation and differential scanning calorimetry measurements. A model has been developed that describes the orientational stability as a function of temperature and processing conditions. The time-temperature relaxation of NLO chromophores in a variety of polymer systems can be understood in terms of current phenomenological descriptions of the glass transition with the aid of a scaling relation.

1. INTRODUCTION

Understanding relaxational processes in nonlinear optical (NLO) polymeric materials is of critical importance in order to evaluate the long-term stability of poled polymers that are in development for potential electro-optic (EO) applications. Insufficient temporal stability of the induced polar ordering of NLO chromophores at elevated temperatures is a major hindrance to further progress in developing polymeric NLO devices. An essential requirement for stabilizing polymeric NLO materials is the formation of a glassy state at relatively high temperatures. Amorphous polymers, among other glasses, typically show evidence of a phase transition from a liquid-like to a glassy state when cooled from high temperatures. The temperature at which this transition occurs, known as the glass transition, T_g, is recognized to be primarily a kinetic phenomenon, whether or not it is valid to classify it as a true phase transition [1]. The physical origin of the glass transition is primarily associated with the cooperative motions of large scale molecular segments of the polymer. The actual experimental observance of a glass transition, however, is most easily probed by measuring enthalpic changes in the polymer as a function of temperature via differential scanning calorimetry (DSC). Several phenomenological theories describe the primary aspects of the glass transition, at least as they are experimentally observed.

We make use of a variation of one of these phenomenological theories in an attempt to gain a better understanding of the temperature dependence of relaxational processes in NLO polymers. An algorithm has been developed that allows us to model most of the observed thermal responses of glassy polymers as they are probed by DSC measurements. The model then enables us to calculate relaxation

---

* *Corresponding author.* Tel.: +41-1-633 23 29. Fax: +41-1-633 10 56.

times as a function of both temperature and processing conditions (cooling rates, annealing times and temperatures, etc.).

The modified polyimide polymers used in this work have moderate to high glass transition temperatures ($140°C < T_g < 190°C$) with the NLO activity arising from side-chain azo chromophores [2]. These polymers offer the possibility of high chromophore densities and concomitant high SHG and EO coefficients. Extensive SHG time-temperature measurements have been done to verify the degree of stability of the induced polar ordering in these polymer films. We have also made corresponding relaxational measurements of the side-chain chromophores above T_g as probed by dielectric relaxation. With the aid of a scaling relation, we find that the relaxation behavior of the NLO chromophores can be reasonably well modeled in terms of current phenomenological descriptions of the glass transition.

The chemical structures of the polymers characterized in this work are shown in Figure 1. The amino-alkyl-amino functionalized NLO azo chromophores, with the various substitution patterns shown in Figure 1, were attached via a two or three carbon spacer linkage to an alternating styrene-maleic-anhydride copolymer. Three polymers denoted by A-095.11, A-097.07, and A-148.02 with the chromophores indicated in Figure 1 were chosen for the detailed measurements described in this work. A discussion of the chemical synthesis is given elsewhere [2].

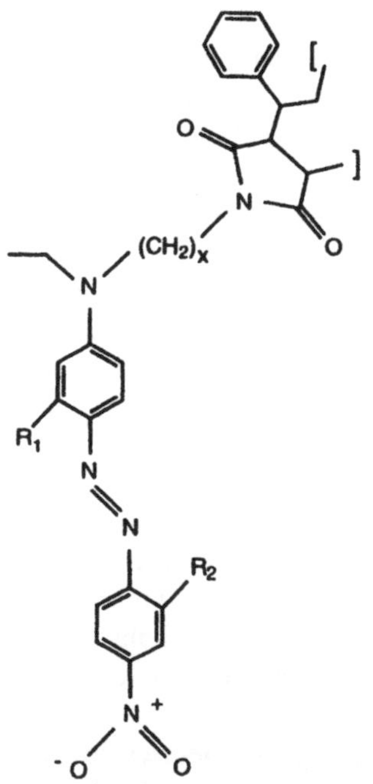

	A-095.11	A-097.07	A-148.02
x	3	3	2
R_1	CH_3	CH_3	H
R_2	H	Cl	H
T_g	137 °C	149 °C	172 °C

Figure 1 Modified NLO side-chain polyimide polymers. The alkyl amino functionalized NLO azo chromophores with the various substitution patterns were attached via a two or three carbon spacer linkage to an alternating styrene-maleic-anhydride copolymer [2].

2. OPTICAL AND NONLINEAR OPTICAL PROPERTIES OF THE POLYMERS

The optical and nonlinear optical properties of the polymers investigated in this study are given in Table 1. The visible and near IR optical properties of the polymers were measured with a Perkin-Elmer $\lambda 9$ UV-VIS-IR spectrometer and nonlinear optical measurements were performed using a standard Maker-fringe technique [3]. Stimulated Raman scattering in compressed H_2 (≈ 40 bar) and CH_4 (≈ 20 bar) in approximately one meter long gas cells provided the fundamental wavelengths of $\lambda = 1907$ nm and $\lambda = 1542$ nm, respectively. The laser could be operated at $\lambda = 1338$ nm by changing the cavity mirrors.

Film Preparation

Films of each polymer with thicknesses ranging from 0.5 to 2.5 μm were spin cast from solutions of cyclopentanone/N-methyl-2-pyrrolidone (NMP) in a mixture ratio of 5:1, with different polymer concentrations, onto ITO-coated glass substrates.

Poling Procedure and Nonlinear Optical Coefficients

Several films of each polymer were poled near the glass transition with a positive or negative corona discharge at voltages between 8 and 15 kV [4]. The distance between the corona needle and the polymer film was approximately 1–2 cm. During the poling process the film was held at a fixed angle of $\theta = 25°$ with respect to the laser beam. After a film was poled, the needle was removed and the film was investigated for its second-harmonic response by rotating the poling stage perpendicularly to the incoming laser beam. The nonlinear optical coefficients were evaluated using the following formula [4]:

$$I_{2\omega}(\theta) = \frac{8}{\varepsilon_0 c} d_{31}^2 [t_{af}^{\omega}(\theta)]^4 \, T_{2\omega}(\theta) \, [t_{sa}^{2\omega}(\theta)]^2 p^2(\theta) I_{\omega}^2 \left(\frac{1}{n_{2\omega}^2 - n_{\omega}^2} \right)^2 \sin^2 \psi, \qquad (1)$$

where $t_{af}^{\omega}(\theta)$ and $t_{sa}^{2\omega}(\theta)$ are transmission coefficients at the air/film and substrate/air interfaces, $T_{2\omega}(\theta)$ is the second-harmonic Fresnel factor, and $p(\theta)$ is a projection factor. The nonlinear optical d coefficients of the polymer films shown in Table 1

Table 1   Linear and nonlinear optical properties of the side-chain polyimide polymers.

Polymer	A-095.11	A-097.07	A-148.02	Poling field
λ_0 [nm]	490	510	470	
n @ 633 nm	1.84 ± 0.01	1.89 ± 0.01	1.80 ± 0.01	
n @ 1313 nm	1.65 ± 0.01	1.67 ± 0.01	1.66 ± 0.01	
d_{31} @ 1338 nm [pm/V]	23 ± 3	23 ± 7	19 ± 5	+8 kV corona
d_{31} @ 1542 nm [pm/V]	15 ± 2	15 ± 2	14 ± 3	+8 kV corona
r_{31} @ 1313 nm [pm/V]	6.7 ± 0.4	2.3 ± 0.2	6.5 ± 0.3	150 V/μm

were calibrated with respect to the SHG signal of a quartz crystal reference ($d_{11} = 0.4$ pm/V). Electric field poling with a positive corona gave NLO susceptibilities approximately twice as high as those obtained from a negative corona [4].

Measurement of the Electro-Optic Coefficient

Either the Michelson interferometric setup [5] or the ellipsometric technique [6] was used to measure the electro-optic coefficient r_{13} of electrode poled films. Films of a thickness of about 1 μm with a 100 nm thick top gold electrode were poled at T_g for about 1/2 hour at poling fields between 120 and 150 V/μm. A diode pumped Nd:YLF laser ($\lambda = 1313$ nm) was slightly focused onto the films and the output was detected with an IR sensitive Germanium photodiode. A modulation voltage $U_{eff} \leq$ 10 V at 1 kHz was applied over the film and used as a reference for a lock-in amplifier. The EO coefficients are obtained from the resulting modulation of the refractive index. Table 1 also gives the results of these measurements [4].

3. MODELING STRUCTURAL RELAXATION IN POLYMERIC GLASSES

When a liquid is subjected to a sudden change in temperature T, its physical properties such as volume V, viscosity η or enthalpy H undergo an instantaneous, solid-like change followed by a slower, liquid-like relaxation to new equilibrium values at the new temperature. It is generally presumed that this structural relaxation involves some change in the average molecular configuration of the liquid. The time $\Delta t'$ required for the relaxation increases rapidly with decreasing temperature. The liquid is said to enter the *glass transition region* when its timescale $\Delta t'$ and the timescale Δt of an observer are of the same order of magnitude. *Above* the glass transition region $\Delta t' \ll \Delta t$ the structure rearranges promptly in response to changes in T and measured properties are said to be those of an equilibrium liquid. *Below* the glass transition region, $\Delta t' \gg \Delta t$, structural rearrangements are kinetically arrested and the material is referred to as a glass.

A useful functional representation for describing the nonexponentiality of structural relaxations in polymeric glasses in the time domain is the stretched exponential or Kohlrausch–Williams–Watts (KWW) function [7]:

$$\phi(t) = \exp - (t/\tau)^b \tag{2}$$

with $0 < b \leq 1$. The issue of nonlinearity of structural relaxations was addressed by Tool by introducing the concept of fictive temperature, T_f, which describes the influence of the actual structure of a glass on the relaxation time τ.

Using an assumption of thermorheological simplicity, Narayanaswamy has given an explicit procedure, which will be denoted the *TN* (*Tool–Narayanaswamy*) procedure, for the quantitative calculation of the fictive temperature [8]. Further details for incorporating nonexponentiality in this procedure have been provided by Moynihan *et al.* [9] by making use of the KWW function in describing the enthalpic relaxation and recovery process. Using this procedure, the value of T_f can be calculated from the previous thermal history by using the following expression:

$$T_f = T_0 + \int_{T_0}^{T} dT' \left\{ 1 - \exp\left[-\left(\int_{T'}^{T} \frac{dT''}{q\tau} \right)^b \right] \right\},$$

(3)

where $q = dT/dt$ is the heating/cooling rate and T_0 is an arbitrary reference temperature above the glass transition temperature.

The structural relaxation time of polymeric liquids, as interpreted by DiMarzio and Gibbs involves the cooperative rearrangement of a number of molecular segments of the polymer molecule [10]. Adams and Gibbs subsequently obtained an expression for the structural relaxation time in terms of the configurational entropy [11]. Hodge has shown that the following expression can be derived for the relaxation time associated with the cooperative chain rearrangements [12]:

$$\tau = A \exp\left(\frac{B}{T(1 - T_2/T_f)} \right),$$

(4)

where $B \cdot R$ is an activation energy with R the gas constant, T is the temperature, and A is a time parameter. We assume that the temperature T_2 in this formulation is equivalent to the "Kauzmann temperature", T_K, which is described by the zero point of the configurational entropy.

Extended DSC (differential scanning calorimetry) measurements of the polyimide NLO polymers indicate the following correlation between the parameters A, B, and T_2 [13]:

$$A = \tau_g \exp\left(\frac{-B}{T_g - T_2} \right) = \tau_g \exp(-2.303 C_1^g),$$

(5)

with τ_g the relaxation time at T_g, typically in the range of 10–100 s. The parameter C_1^g is one of the Williams–Landel–Ferry (WLF) parameters with a "universal" value of about 17 [14].

Above the glass transition the polymer system is in thermal equilibrium and therefore, T_f equals T. Together with the replacement of A in Eq. (4), this yields the functional form of the Vogel–Fulcher (VF) equation for the normalized relaxation time, $\tau_n = \tau/\tau_g$:

$$\ln \tau_n = -\frac{B}{T_g - T_2} + \frac{B}{T - T_2}.$$

(6)

Below the glass transition T_f reaches a final limiting value in simple cooling processes, i.e. without sub-T_g annealing. Assuming the final fictive temperature to be T_g, we obtain the following temperature dependence of the normalized relaxation time:

$$\ln \tau_n \approx \frac{B}{T_g - T_2} \frac{T_g - T}{T} = 2.303 \, C_1^g \frac{T_g - T}{T} \tag{7}$$

This result indicates that the primary or α-relaxation processes should have an Arrhenius temperature dependence below the glass transition. In addition, from the "universality" of the (WLF) parameter, C_1^g, one might expect normalized relaxation times to scale with the scaling parameter $(T_g - T)/T$, at least in the same class of glass forming liquids.

4. DIFFERENTIAL SCANNING CALORIMETRY (DSC) MEASUREMENTS

DSC measurements were done with a commercial Perkin Elmer DSC-2C apparatus. The samples were prepared as pressed powder capsules of about 15–30 mg [13]. A typical measurement procedure was accomplished as follows: the polymer sample was first heated to an initial temperature of about $T_g + 50$ K for approximately 15 min. Data was collected on the heating portion of the thermal cycle as a function of cooling rate and with a constant heating rate of 20 K/min. The measured heat flow $\mathrm{d}Q(T)/\mathrm{d}t$ is related to the polymer heat capacity $C_p(T)$ by

$$\frac{\mathrm{d}Q(T)}{\mathrm{d}t} = m_p q \, C_p(T), \tag{8}$$

where m_p is the polymer mass and $q = \mathrm{d}T/\mathrm{d}t$ is the heating/cooling rate. The calculated heat capacity can be compared to the DSC measurement by defining the normalized heat capacity

$$C_P^N(T) = \frac{\mathrm{d}T_f}{\mathrm{d}T} = \frac{C_p(T) - C_{pg}(T)}{C_{pe}(T) - C_{pg}(T)}, \tag{9}$$

where T_f is given by Eq. (3). C_{pg} and C_{pe} are the heat capacities in the glassy and equilibrium temperature regions sufficiently extrapolated from the relaxation temperature region.

An extensive set of DSC traces for A-148.02 is shown in Figure 2. The curves with the lowest cooling rates were taken at the end of the series. The shift in peak temperature might be due to chemical degradation of the chromophores, since the same sample was used for the series of DSC measurements. Theoretical expectations based on the parameter set with $q = -5/20$ K/min are displayed on the left side. As seen in Figure 2, the calculated peak height and peak position are in good agreement with the experimental results. However, the overall peak shape demands different parameter sets to obtain best fits. The parameters for the overall best fits shown in

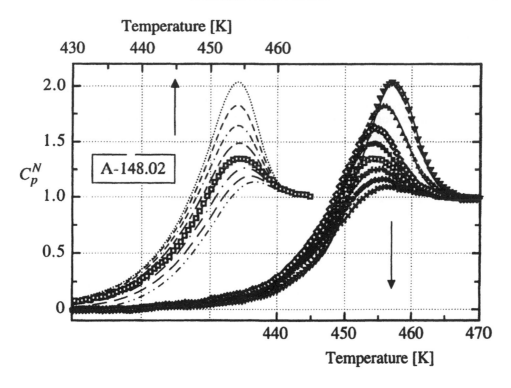

Figure 2 Normalized C_p^N DSC heat capacity curves of the polyimide side chain polymer A-148.02 for different cooling rates ranging from -0.31 to -40 K/min. The heating rates were all at 20 K/min. The right side displays experimental results with best fits. The left side gives the theoretical expectation based on the parameter set of the trace (open squares) with a cooling rate of 5 K/min.

Table 2 Enthalpic Adam–Gibbs parameters for the polymers investigated in this work.

Polymer	B [10^4 K]	T_2 [K]	T_g [K]	b	$\log(\tau_g[\text{s}]/\text{s})$
A-095.11	2.5 ± 0.4	160 ± 20	410	0.45 ± 0.04	1.7 ± 1.0
A-097.07	6 ± 2	160 ± 20	422	0.25 ± 0.06	2.5 ± 0.8
A-148.02	2.7 ± 0.5	170 ± 40	444	0.50 ± 0.03	1.2 ± 0.6

Figure 2 are given in Table 2 [13]. The computation of enthalpic relaxation processes as measured by DSC was done by numerical integration of Eq. (3) using Eq. (9). Parameter optimization was accomplished via a Levenberg–Marquardt fitting algorithm.

5. DIELECTRIC RELAXATION MEASUREMENTS
ABOVE THE GLASS TRANSITION

A Hewlett–Packard HP 4129A LF Impedance Analyzer with a frequency range of
5 Hz to 13 MHz was used for dielectric measurements to investigate the relaxation of
the side-chain chromophores above the glass transition temperature. The impedance
measurements were done according to the "lumped circuit" method in either a
parallel or series configuration. Typical values of the capacitance of the films were in
the range 0.1–1.0 nF [13].

All of the dielectric loss spectral measurements show a large rise at low frequencies. The large increase in ε'' is most likely due to the DC conductivity σ_0 of the
polymer which is usually described as follows:

$$\varepsilon''_{DC}(\omega) = \frac{\sigma_0}{\varepsilon_0 \omega}. \tag{10}$$

A common method of treating the DC conductivity of the samples is to fit it with
a potential function of the form $a\omega^k$ with $k \approx -1$ and subtract it from the measured
data. With high T_g polymers, however, this procedure is not suitable, since the
ε''-resonance merges with the DC conductivity in ε'' at low frequencies, and is seen
just as a shoulder rather than as a peak; see Figure 3. Instead, we use an analytical

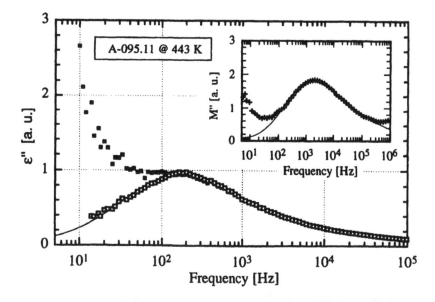

Figure 3 Correction procedure to obtain the dielectric loss relaxation times. The polymer
conductivity prevents an accurate fit to ε''. Use of the electric modulus M'', however, allows
the dielectric loss peak to be deconvolved from the DC conductivity (see inset).

scheme developed by Howell *et al.* [15]. By analogy to mechanical relaxation, an 'electric modulus' M^* is defined, which may be termed a complex inverse permittivity:

$$M^*(\omega) \equiv 1/\varepsilon^*(\omega) = (M_\infty - M_0)[1 - N^*(\omega)] + M_0, \qquad (11)$$

where $N^*(\omega) = N'(\omega) - iN''(\omega)$. M_∞ is the high and M_0 the low frequency limit of the real part of M^*. First, the imaginary part M'' was calculated from measured ε^*-data and subsequently fitted according to Eq. (11). Secondly, the results for M'' together with the real part data M' from the first conversion were backconverted to yield corrected ε''-values. Finally, these data were again fitted to the Laplace transform of the KWW decay function, Eq. (2). An illustration of this correction procedure is shown in Figure 3.

The temperature dependence of the measured relaxation times was well described by either the VF or WLF expressions [13]. The VF equation, Eq. (6) is related to the WLF equation which is usually written as [14]

$$\log \tau_n = -\frac{C_1^g(T - T_g)}{C_2^g + (T - T_g)} = -C_1^g + \frac{C_1^g C_2^g}{T - T_2}, \qquad (12)$$

where B of the VF equation is related to the WLF parameters by $B = 2.303\, C_1^g C_2^g$. The temperature T_2 can be written as $T_2 = T_g - C_2^g$, with $C_2^g \approx 50°C$ if "universal" relaxational behavior (in the WLF sense) is assumed. The measured values of the WLF parameters are close to these universal values. Table 3 gives the results of the dielectric measurements above the glass transition with the best fit parameters to Eqs. (6) and (12) [13].

6. SECOND-HARMONIC (SHG) DECAY MEASUREMENTS BELOW THE GLASS TRANSITION

Relaxation of the side-chain chromophores below the glass transition was investigated by the decay of the SHG signal from either corona or electrode poled films. For the electrode poled films, a gold film was deposited onto the polymer film to provide a second electrode. The polymer films were baked at approximately 170°C for one day to remove any residual solvent. After poling slightly above the glass transition temperature, the polymer films were cooled at a rate of $\approx$ 2–5°C/min to

Table 3 Vogel–Fulcher and WLF parameters from dielectric relaxation.

Polymer	B [10^3 K]	T_g [K]	T_2 [K]	C_1^g	C_1^g	$\log(\tau_g$ [s]/1s)
A-095.11	1.5 ± 0.3	410	356 ± 9	12 ± 3	54 ± 9	1.3 ± 0.6
A-097.07	1.2 ± 0.2	422	377 ± 7	12 ± 2	45 ± 7	2.1 ± 0.3
A-148.02	1.5 ± 0.4	445	389 ± 15	11 ± 4	56 ± 15	1.0 ± 0.3

the decay temperature which was chosen to be in the range from 80°C up to T_g for the individual polymers. Either a 30 Hz Q-switched Nd:YAG laser operating at 1338 nm or a 800 Hz Q-switched Nd:YLF laser operating at 1313 nm was used for the SHG decay experiments.

Corona Poled Films

At the decay temperature the corona field was turned off and the SHG signal was continuously monitored for a time period of up to 100 hours depending on the decay temperature. SHG decay at 80°C was monitored for more than one year, with data points being measured in logarithmic time intervals.

Electrode Poled Films

At the decay temperature the electrodes were shortcircuited and the SHG signal was continuously monitored for a time period of up to 100 hours depending on the decay temperature.

In both cases the SHG signal was detected with a PMT and a boxcar averager with 0.1–1 s integration time and active baseline subtraction. A quartz reference signal was monitored concurrently to reduce laser fluctuations and correct for drifts in the laser intensity level over the acquisition time of the decay measurements. A series of normalized decay traces for A-095.11 is shown in Figure 4. The

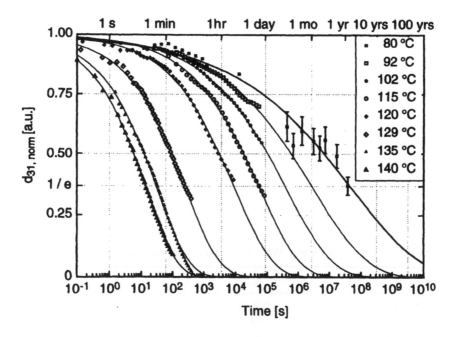

Figure 4   SHG relaxation traces for A-095.11 for different decay temperatures.

fit is to the KWW function, Eq. (2), after a correction with an exponential term describing the influence of the electric field induced third-order effect, and the release of charge carriers trapped in the film in the case of corona poling [13].

7. SCALING OF RELAXATION TIMES FOR NONLINEAR OPTICAL POLYMERS

To describe the global relaxation behavior, we used the *TN* model with parameter values taken from Table 3 using the WLF fits [13]. Rewriting Eq. (4) in terms of the WLF parameters gives

$$\tau_n = \exp\left(-2.303 \cdot C_1^g + \frac{2.303 \cdot C_1^g C_2^g}{T(1 - T_2/T_f)}\right), \tag{13}$$

where T_f is calculated by Eq. (3). With a cooling rate of 5 K/min (corresponding to the cooling rate used in SHG relaxation experiments) we calculate relaxation times as indicated by the solid line in Figure 5. The only adjustable parameter $b = 0.21$

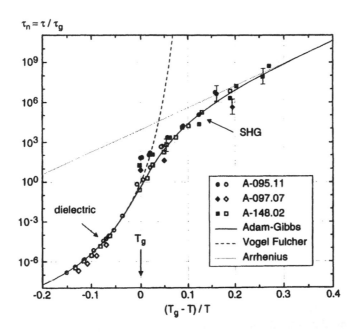

Figure 5 Scaling of normalized relaxation times for NLO side-chain polymers. Filled points belong to corona poled, open points to electrode poled films. The solid line is calculated according to the Tool–Narayanaswamy procedure based on the Adam–Gibbs model for the relaxation time τ. Input parameters are taken from dielectric measurements. The Vogel–Fulcher law fitted to dielectric data is given by the dashed line. Far below the glass transition, the relaxation is of Arrhenius type with an activation energy of $1.3 \cdot 10^5$ J/mol.

was determined by a least squares fit (on a logarithmic time scale) to all the data from the electrode poled films. This value is close to b from SHG relaxation at the lowest temperatures.

Relaxation times from dielectric and SHG decay measurements are shown in Figure 5. The results show that it is possible to model the relaxational behavior of NLO chromophores both above and below the glass transition over more than 15 orders of magnitude in time using the *TN* procedure by incorporating the appropriate WLF parameters for the NLO polymer. This procedure including our minor modification provides an excellent framework for the nonlinear extension of the liquid state relaxation behavior through the glass transition as suggested by the consistency between observed and calculated relaxation times below the glass transition.

8. CONCLUSIONS

As shown in Figure 5, the dielectric and SHG relaxation results are well explained by the *TN* formalism incorporating the WLF parameterization. An intuitive understanding of these results are best clarified through the use of free-volume concepts. The WLF parameters C_1^g and C_2^g can then be expressed in terms of free-volume parameters as follows [14]:

$$C_1^g = \frac{1}{2.303} \frac{\bar{b}}{f_g}, \qquad C_2^g = \frac{f_g}{\alpha_f}, \tag{14}$$

where f_g is the fractional free volume at T_g, α_f is the coefficient of thermal expansion and $\bar{b}$ is an empirical material constant. Thus, the slope of our "universal" relation, Eq. (7), describing relaxation below T_g is inversely proportional to the fractional free volume available to the chromophores at the glass transition temperature. Although the WLF parameter C_1^g is known to be polymer dependent, values of $f_g/\bar{b}$ determined by mechanical measurements vary over a relatively small range from about 0.013–0.034. This implies that the values of C_1^g range from approximately 13–34. From SHG decay measurements on a variety of polymeric systems, we find that the slope of Eq. (7) (given by C_1^g) falls near the lower end of this range. Thus, for NLO polymers that incorporate chromophores of approximately the same size we would expect scaling in a plot of relaxation times *vs.* $(T_g - T)/T$ to be "universal" in the sense that the fractional free volume probed by the chromophores should be nearly the same.

The algorithm developed in this work (Eq. (13)), enables us to calculate relaxation times as a function of both temperature and processing conditions. As an example, we can apply these results to a hypothetical EO device utilizing a NLO polymer as the active material. In order to limit the decrease in the EO coefficient to less than 5% of the initial value over a device lifetime of 5 years and operation at 80°C, a glass transition temperature of about 270°C is needed ($\tau = 2.2 \times 10^{14}$ s, $(T_g - T)/T = 0.53$, $b = 0.21$). The polymer A-148.02 would retain only 56% of the initial nonlinearity under these conditions. NLO polymers having glass transition temperatures in this range and higher are under current research development [16].

REFERENCES

[1]   McKenna, G. B., in *Comprehensive Polymer Science*, Vol. 2, G. Allen, ed. (Pergamon, New York, 1989) 311.
[2]   Ahlheim, M. and Lehr, F., *Makromol. Chem.* **195** (1994) 361-373.
[3]   Jerphagnon, J. and Kurtz, S. K., *J. Appl. Phys.* **41** (1970) 1667.
[4]   Prêtre, P., Kaatz, P., Bohren, A., Günter, P., Zysset, B., Ahlheim, M., Stähelin, M. and Lehr, F., *Macromolecules* **27** (1994) 5476-5486.
[5]   Bosshard, C., Sutter, K., Schlesser, R. and Günter, P., *J. Opt. Soc. Am. B* **10** (1993) 867.
[6]   Teng, C. C. and Man, T. H., *Appl. Phys. Lett.* **56** (1990) 1734-1736.
[7]   Williams, G., Watts, D. C., S.B., D. and North, A. M., *Trans. Far. Soc.* **67** (1971) 1323.
[8]   Narayanaswamy, O. S., *J. Am. Cer. Soc.* **54** (1971) 471.
[9]   Moynihan, C. T., Crichton, S. N. and Opalka, S. M., *J. Non-Cryst. Solids* **420** (1991) 131-133.
[10] DiMarzio, E. A. and Gibbs, J. H., *J. Chem. Phys.* **28** (1958) 373.
[11] Adam, G. and Gibbs, J. H., *J. Chem. Phys.* **43** (1965) 139-145.
[12] Hodge, I. M., *Macromolecules* **19** (1986) 936.
[13] Kaatz, P., Prêtre, P., Meier, U., Stalder, U., Bosshard, C., Günter, P., Zysset, B., Ahlheim, M., Stähelin, M. and Lehr, F., *Macromolecules*, **29** (1996) 1666-1678.
[14] Ferry, J. D., *Viscoelastic Properties of Polymers* (J. Wiley & Sons, New York, 1980).
[15] Howell, F. S., Bose, R. A., Macedo, P. B. and Moynihan, C. T., *J. Phys. Chem.* **78** (1974) 639-648.
[16] Miller, R. D., Burland, D. M., Jurich, M., Lee, V. Y., Moylan, C. R., Thackara, J. I., Twieg, R. J., Verbiest, T. and Volksen, W., *Macromolecules* **28** (1995) 4970-4974.

9. RELAXATION AND AGING IN POLED POLYMER FILMS

H.-J. WINKELHAHN, D. NEHER*, Th. K. SERVAY[1], H. RENGEL,
M. PFAADT, C. BÖFFEL, and M. SCHULZE[2]

*Max-Planck-Institute for Polymer Research, Ackermannweg 10,
D-55128 Mainz, Germany*

ABSTRACT

The orientational relaxation kinetics is analyzed by isothermal decay experiments and stimulated current measurements for side chain EO-polymers with different substituents. The polar relaxation does not obey thermorheological simplicity but shows a strong broadening of the relaxation time distribution in the glassy state. The temperature dependence of relaxation times as determined by dielectric spectroscopy and polar decay measurements could be described consistently by the Adams–Gibbs equation. Aging effects are present in most compounds as shown by a strong correlation between the relaxation behavior and the thermal history of the samples.

INTRODUCTION

The temporal stability of the electrooptical response in one of the key performance parameters in the assessment of poled polymers for the application in electric field controlled light modulators or directional couplers. Even though a universal definition of temporal stability has not been given, life times of the device of approximately 5–10 years at elevated working temperatures are required. The change in device performance with time will be due to different processes such as photoinduced chemical degradation or the decrease in the electrooptical coefficient caused by the relaxation of the polar order of the EO-active chromophores. In particular the latter effect has been studied by various groups [1–14].

Different methods have been employed to study relaxation processes in glass forming polymers. NMR spectroscopy is certainly the most powerful method since it allows the assignment of a particular macroscopically observed relaxation process to the motion of a specific chemical moiety. However, this method is limited to correlation times below approximately one second [15]. Dielectric spectroscopy is sensitive to the motion of molecular sub-units bearing a permanent dipole moment. Again, the frequency range of this method, with maximum relaxation times of about 1000 s, and contributions from ionic impurities often do not allow the investigation of very slow relaxation processes. Poled polymers have thus mostly been investigated utilizing isothermal experiments, in which either the second harmonic signal or the electrooptical response is recorded as a function of time for a fixed temperature. Values for the polar relaxation times are often given in connection to a particular

---

* *Corresponding author.*
[1] Current address: BASF AG, 67056 Ludwigshafen, Germany.
[2] Current address: Royal, Institute of Technology, Department of Polymer Technology,
S-100 44 Stockholm, Sweden.

relaxation time distribution such as the discrete multiexponential decay or the KWW stretched exponential function.

Time–temperature characteristics have been explained by the WLF equation [3] and Arrhenius-like behavior [2,6,16], as well as by empirical equations [1,14,17] depending on the type of chromophore and its linkage to the polymer chain. The validity of these interpretations can only be manifested by the analysis of the relaxation process over a wide temperature range. Since rotational motion occurs in the time range of equal or less than 100 s above the glass transition temperature, poled polymers must be in the glassy state in order to show sufficient long term stability. Therefore, the relaxation properties will depend not only on the actual temperature but also correlate with parameters describing the glassy state.

The goal of this paper is to summarize and discuss our recent investigations on the polar order decay for different EO-active polymers with emphasis on the correlation between the relaxation dynamics and the properties of the glassy state. We will show, based on experimental results and theoretical models, that non-Arrhenius-type activation should be generally expected for the polar relaxation of glass forming polymers.

MATERIALS

The structures of the different polymers under study are shown in Figure 1. PB ($T_g = 72°C$) is based on a flexible PMMA-type main chain [18]. PC25 [19,20] ($T_g = 40°C$), PC25a [21] ($T_g = 38°C$), PC27 [22] ($T_g = 32°C$) and PC57 [19] ($T_g = 42°C$) consist of a rigid rod-like polyester main chain substituted with aliphatic side chains of different length and an EO-active chromophore linked to the backbone by a short flexible spacer. In PC4 ($T_g = 87°C$) and PC6 ($T_g = 68°C$) [20] some of the aliphatic substituents have been replaced by benzyloxy-groups resulting in an increase in T_g. Finally PCk ($T_g = 42°C$) is based on a semi-flexible backbone [23]. All compounds with a rigid backbone structure form a microphase-separated structure with layers of the stiff main chains separated by side chain layers. DSC, dielectric spectroscopy and X-ray diffraction experiments showed that the EO-active chromophore is embedded in the side chain region. The glass transition, which had not been observed for polyesters substituted exclusively with aliphatic side chains, does not affect the order within the main chain layer but corresponds to an increased mobility of the side chains region. Despite the presence of the stiff polymer backbone in most of these compounds the glass transition temperature ranges only between 40°C and 90°C.

EXPERIMENTAL

Sample Preparation

Thin film samples for the thermally stimulated discharge current and waveguide experiments were spin coated from a polymer/tetrachloroethane solution onto a 40 nm thick gold layer deposited on a BK7 glass substrate. Spinning conditions were

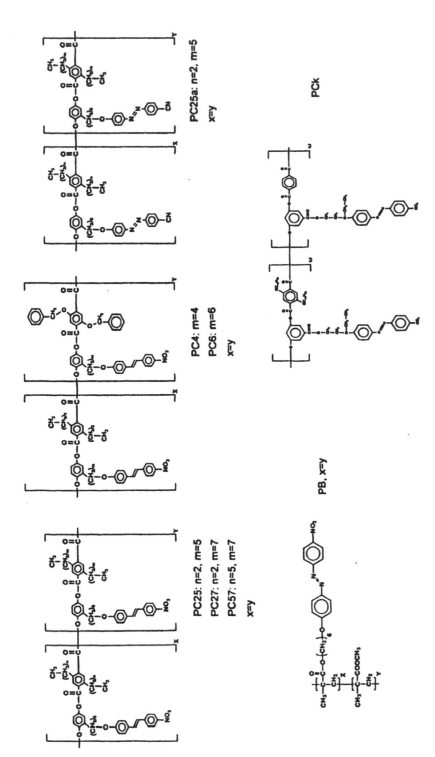

Figure 1 EO-active side chain polymers studied in this work.

chosen to obtain a film thickness of about $2\,\mu m$. After the film have been dried for 24–48 h in a vacuum oven, a 100 nm thick top gold electrode was evaporated on top of the polymer layer. Electric field poling was performed at T_{pol} above T_g with variable poling fields. The poling temperature was chosen such that the dynamical and orientational properties of the material are in thermal equilibrium and the previous thermal history is extinguished. Poling and relaxation studies were carried out in nitrogen atmosphere to prevent any influence of air humidity on the polar decay characteristics. Following the commonly used poling procedure, linear cooling with rates q_c varying between 0.1 and 10 K/min were used from T_{pol} to room temperature with the poling field still applied.

Isothermal Decay Measurements

Isothermal relaxation experiments have been carried out using the method of Attenuated Total Reflection (ATR) [17] or with an interferometrical set-up [24]. Measurements started immediately after the samples were re-heated from RT to the actual relaxation temperature and the poling field had been switched off.

Thermally Stimulated Discharge Currents (TSDC)

The TSDC setup used in these investigations has been described in detail in [25]. After cooling down to room temperature with the poling field still applied, the field was switched off and the samples were re-heated at a constant heating rate of 10 K/min. The resulting depolarization current was measured using a Keithley 485 picoamperemeter.

DSC

The DSC traces have been recorded using a Mettler DSC 30. We used the same thermal conditions as for the TSDC experiments, namely a heating rate of 10 K/min and cooling rates varying between 0.1 and 10 K/min.

Dielectric Spectroscopy

Samples for dielectric spectroscopy have been prepared by melting the polymer in vacuum between two nickel plated brass electrodes. The film thickness was controlled by a $50\,\mu m$ glass fiber inserted as spacers. The dielectric properties have been determined by a Solatron–Schlumberger SI-1260 in the frequency interval from 10^{-2} to 10^6 Hz. The temperature of the samples was controlled by a heated nitrogen gas jet.

RESULTS

Effect of Poling Field Strength

Samples have been poled well above T_g with different strengths of the applied poling field from 1.1 to 54.5 V/μm. Since the EO-response of samples poled with low fields was rather weak, we used the sensitive and background free TSDC method. Figure 2 shows the corresponding TSDC scans, normalized for the electric field strengths. As observed by others [5], we detect two different processes: the α-peak appearing at lower temperatures, which can be connected with the glass transition of the polymer and a second peak attributed to the motion of space charges induced during the poling process (ρ-peak). There are no indications that either the position or the shape of the α-peak is influenced by the electric field strength. Irreproducible fluctuations were, however, observed at higher fields (ρ-peak). Since the polar relaxation has been shown to correlate to the α-peak, only, [25] the isothermal relaxation behavior is expected to be independent of the magnitude of the electric field. Field dependent effects have however been observed by Wang and co-workers

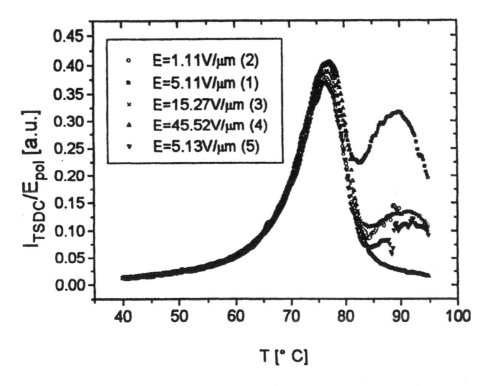

Figure 2 TSDC scans of a sample of polymer PB, which had been poled with different poling fields E_{pol}.

for larger electric fields between 80 and 190 V/µm, well above the field strengths used in our experiments [9].

Relaxation Time Distributions

Figure 3 shows the isothermal decay of the electrooptical signal as function of time for different temperatures, normalised for the response directly after switching-off the poling field. The decay curves plotted as function of log (*t*) cannot be overlapped by a horizontal shift of the graphs, indicating non-thermorheological-simplicity (non-TRS). TRS behavior, also called temperature-time superposition or principle of

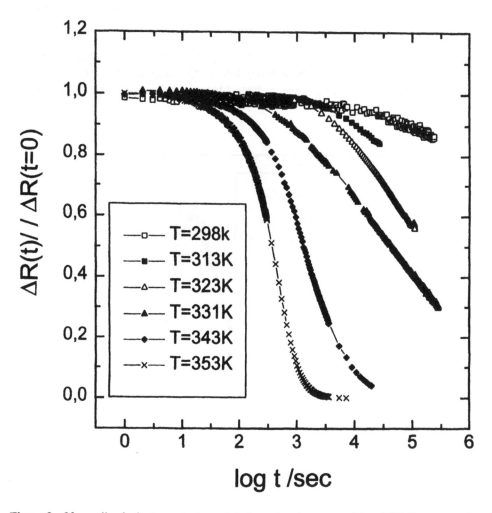

Figure 3 Normalized electrooptical modulation signal measured by ATR for a sample of polymer PC6 for different relaxation temperatures plotted as function of log(*t*). A cooling rate of 0.1°C/min had been used in the poling procedure.

reduced variables, has often been observed in glass forming polymer systems above T_g and is one of the main requirements for the construction of so-called master curves over large time or temperature ranges [26]. TRS means that each relaxation time within the relaxation time distribution has the same temperature dependence. Our data indicate that a significant change of the relaxation time distribution with temperature must be present in our poled polymer system below T_g.

The experimental data could be fitted either to a bi-exponential decay and by a KWW-stretched exponential function. Superior agreement to the experimental data could be obtained assuming a symmetric Havriliak–Negami (HN) relaxation time distribution with the fitting parameters τ_{HN} (mean relaxation time), α and γ, describing the width and asymmetry of the distribution [25,27]. In all cases we found γ close to 1 indicating a symmetric relaxation time distribution. Within the experimental uncertainty the width of the distribution increases with decreasing temperature. This agrees with observations on guest–host systems where best fits to the isothermal decay data were obtained by using a symmetric Gaussian distribution of activation energies [6]. Again, an increase in the width of the distribution with decreasing temperature had been observed.

It is important to recognize that relaxation times deduced from isothermal experiments depend significantly on the choice of the particular relaxation time distribution used to fit the data [27]. Strong derivations can be expected in particular for broad distributions. For example the bi-exponential fit to the isothermal decay of PC6 at 319 K yielded relaxation times of 4.4×10^3 s and 4.0×10^5 s, the KWW function $\tau_{KWW} = 1.3 \times 10^5$ s and the symmetric Havriliak–Negami distribution a mean relaxation time of 6.6×10^4 s. Rather consistent values could be given only by calculating the first logarithmic moments of the relaxation time distributions [27].

A multiexponential decay could be excluded for all side chain polymers studied based on the detailed analysis of the α-peak in the TSDC scan (Figure 4). It is noteworthy to mention that the analysis of the rotational diffusion equation in three and also in two dimensions predicts an almost mono-exponential decay with relaxation times of 1/2D and 1/D for the three and two dimensional case, respectively [13,16]. The nonexponential decay behavior observed thus requires the analysis of complex physical models, which is beyond the scope of the present paper.

Relaxation Kinetics Below and Above T_g

As outlined in the introducing chapter, orientation relaxation above T_g with relaxation times below 100 s can be analyzed using the dielectric spectroscopy. Slow sub-T_g dynamics are usually monitored by isothermal experiments. If the rotational motion of the EO-active chromophore can be well separated in the dielectric spectra, dielectric spectroscopy and polar relaxation experiments should give consistent results over a large temperature and frequency range. This has been nicely demonstrated by recent work of Torkelson and co-workers [2,4]. However, strong derivations have been frequently observed [6,17,23]. At this point one should realize that measurements above T_g are performed with the sample in thermal equilibrium while most polar decay experiments are performed below the glass transition

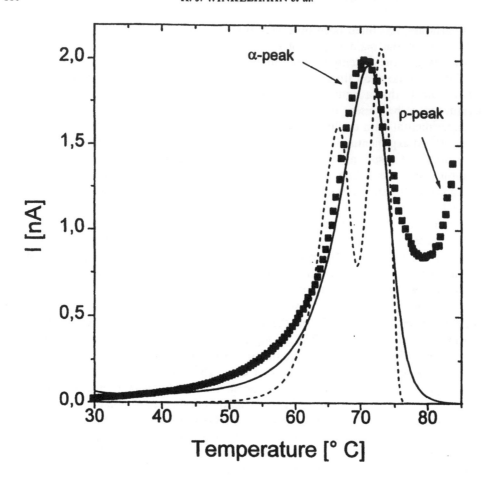

Figure 4 Experimental (solid squares) and theoretical TSDC spectra for PC6 [25]. Theoretical spectra have been computed using a discrete bi-exponential (dashed line) and a continuous Havriliak–Negami (solid line) relaxation time distribution. Corresponding relaxation parameters were deduced from the fits to the isothermal relaxation data.

temperature in the nonequilibrium glassy state, after the sample has been cooled down from the poling temperature. Since various properties of the polymer in the glassy state depend on the thermal history of the sample, the particular thermal treatment should be reflected in the relaxation behavior of the sample.

This effect is demonstrated in Figure 5 for a sample of PC6 [27]. The lower part shows the results of three DSC scans, the upper part the TSDC spectra with corresponding thermal histories. In both experiments the samples have been cooled with cooling rates of 0.1, 1 and 10 K/min, respectively. After each scan the samples have been heated to a temperature 40° above the glass transition in order to erase the former thermal history. The heating cycle was performed with a rate of 10 K/min, consistent with the heating rates involved in the thermal procedure of the isothermal experiments.

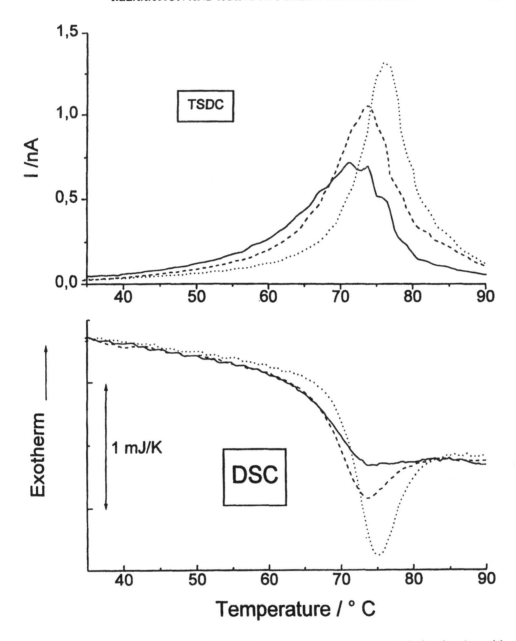

Figure 5 TSDC (upper curves) and DSC (lower curves) scans recorded during heating with a heating rate $q_h = 10$ K/min for polymer PC6. Previously, the samples have been cooled down from a temperature 40 K above T_g with cooling rates q_c of 10 K/min (solid line), 1 K/min (dashed line) and 0.1 K/min (dotted line) [27].

Both structural relaxation and angular redistribution do react in the same way to changes in the cooling rate q_c. With decreasing q_c the DSC spectra show an increasing endothermic peak (the well known DSC overshot) which is related to the glass transition. Slower cooling shifts the glass transition to higher temperatures. The increase in the endothermic signal is solely due to the nonexponential and nonlinear behavior of the structural relaxation and might be modeled by the iteration scheme based on Narayanaswamy's ideas [28]. At the same time the maximum of the α-peak in the TSDC spectra shifts to higher temperatures accompanied by a narrowing of the peak with decreasing cooling rate (see also Ref. [5]). Since the peak position is given by the temperature where the mean relaxation time approaches values between minutes and seconds, close to T_g, a shift in the glass transition temperature with decreasing cooling rate thus causes an increase in the TSDC peak temperature.

A correlation between the temperature dependence of relaxation times and the effect of the thermal history on the thermal properties of the glass forming polymer can be achieved by applying the widely accepted concept of a fictive temperature, originally introduced by Tool to explain structural relaxation in glass forming polymers [29].

This idea has been the basis of several extensions to model the nonequilibrium properties from the knowledge of equilibrium properties [30]. In this approach the actual value of a thermodynamical property P at a given temperature T is given by

$$P(t,T) - P(0,T_o) = [\alpha_p^l(T_f(t,T) - T_o) + \alpha_P^g(T - T_f)]. \tag{1}$$

$P(0,T_o)$ is the value of P at a temperature T_o in the equilibrium liquid state and $\alpha_p^{g,l} = \partial P/\partial T \, (t \to \infty)$ are the slopes of the equilibrium lines in the glassy and liquid state, respectively. This scheme can be easily understood if one considers the temperature dependence of the free volume V_f in the glassy state. The amount of V_f at a given temperature T after a defined cooling cycle starting at T_o above T_g can also be obtained by cooling the sample slowly down along the liquid equilibrium line to T_f followed by very fast quenching to T which maintains the actual amount of free volume. T_f is called the fictive temperature. T_f itself depends strongly on the thermal history of the sample [31] and will, due to aging of the sample, decrease with time to finally approach the actual temperature T. The fictive temperature is thus a measure for the deviation from equilibrium properties. For a given thermal treatment T_f can be calculated by the multiparameter approach by Kovacs, Aklonis, Ramos and Hutchinson [32] or by an iteration scheme developed by Narayanaswamy [28]. While Narayanaswamy had to use empirical parameters to account for the different influences of actual and fictive temperature on the relaxation properties, it was Scherer [33] who extended the Adam–Gibbs [34] theory by the contribution of the fictive temperature. This leads to the following equation for the time–temperture behavior of relaxation times:

$$\tau(T, T_f(T)) = \tau_o \exp\left[\frac{B}{T/T_f(T)(T_f(T) - T_2)}\right]. \qquad \text{AGV} \tag{2}$$

The label AGV was chosen, because Equation 2 is based on the Adam–Gibbs (AG) theory and since it leads to the Vogel-Fulcher (V) equation in the equilibrium state, which is defined by $T_f(T) = T$. The AGV equation has been used by Hodge to simulate the structural relaxation observed in DSC experiments [31].

If P is the total volume of the sample, the amount of free volume will be proportional to $T_f - T_2$. Above T_g, in thermal equilibrium, T and T_f are equal and the relaxation times scale directly with $T_f - T_2$. For fixed activation parameters τ_0, B and T_2, $\tau(T)$ thus solely depends on the amount of free volume. This corresponds to the Doolittle model [35,36]. Below T_g the relaxation rate depends not only on $T_f - T_2$, but also directly on the actual temperature T. For very low temperatures, T_f will be almost independent of T and the AGV equation predicts thermally activated Arrhenius-type behavior.

If $P(T)$ is the enthalpy $H(T)$ of the system, the following relation for T_f can be derived from (1)

$$\frac{dT_f(T)}{dT} = \frac{C_p(T) - C_p^g(T)}{C_p^l(T) - C_p^g(T)} = C_p^f(T), \tag{3}$$

C_p is the specific heat at the actual temperature T and C_p^g and C_p^l are the values for the glassy and liquid state far away from T_g. Therefore, the temperature dependence of relaxation times can be modeled by combining the results of the dielectric spectroscopy, yielding the activation parameters τ_0, B and T_2, and the temperature dependence of the specific head as deduced from the DSC experiment to give $T_f(T)$. This is demonstrated for the example of PC6 in Figure 6. In fact, the relaxation times of polar relaxation in the glassy state can be well predicted from the activation behavior of the rotational motion of the chromophore in equilibrium and the parametrisation of the thermal properties via T_f. The AGV concept could also be used successfully to describe the polar relaxation of polymer P4 [37]. A similar approach, based on the AGV equation, was applied to orientation dynamic data of polyimide side chain polymers [38]. We therefore presume that this approach is generally applicable to glass forming materials.

A different situation was encountered in the investigation of the side chain polymers PC25 [39] and PCK [23]. Figure 7 compares the DSC traces for PC25 for three different cooling rates. For the slowest cooling a broad endothermic peak appears above T_g indicating that considerable crystallization must have occurred during the cooling process. As a consequence the relaxation times determined by dielectric spectroscopy, for which the samples had been cooled rather quickly from the melting temperature down to RT and the results of the polar relaxation experiments with the sample poled and cooled down at a slow rate of 0.1°C/min cannot be combined in a meaningful picture (Figure 8). It remains unclear why crystallization effects did only occur for polymer PC25 and PCK. In both cases the main chains are substituted with rather short alkyl chains. Certainly, the presence of the stiff main chains increases the tendency to crystallize.

The degree of aging for different polymers can be determined quite easily by relating TSDC scans, performed on aged samples, to the glass transition temperature as

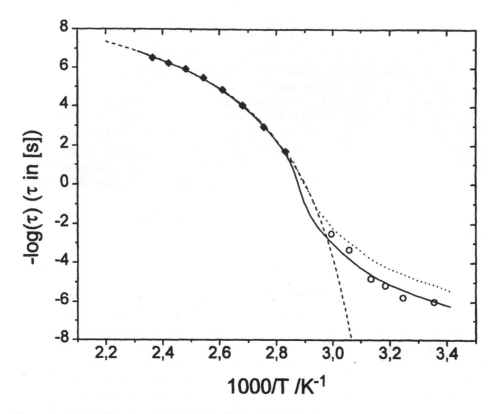

Figure 6 Arrhenius type presentation of the relaxation times as determined from dielectric spectroscopy and isothermal polar relaxation experiments [27]. Diamonds represent relaxation times taken from the Havriliak–Negami fits to the dielectric data. Circles show mean relaxation times deduced from the HN fit to the isothermal decay experiments. During poling, the samples had been cooled down from the poling temperature 10 K above T_g to RT with a cooling rate of 0.1 K/min. The dashed line is the fit of the AGV equation (2) with $T_f = T$ to the dielectric spectroscopy data, yielding the activation parameters τ_o, B and T_2. The solid line was calculated from Eq. (2) in connection with the $T_f(T)$ dependence deduced from the DSC thermograms for a cooling rate q_c of 0.1 K/min (Figure 5). The dotted line shows a simulation assuming $q_c = 1$ K/min.

determined by DSC on samples, using the commonly used cooling and heating rate of 10°C/min. This is demonstrated in Figure 9, for all polymers in Figure 1. Strong shifts of the current peak maximum are indeed observed for compounds PC25, PC25*a* and PC*k*. Only minor aging effects are present for PC4, PC6 and PC57 and for PB with a flexible polymer backbone. Obviously the tendency to age or even crystallize depends on the particular side chain substitution pattern. Smallest effects are observed for those polymers, in which three different types of substituents, namely the EO-chromophore, a flexible alkyl chain and an aromatic phenoxy-group have been used.

As shown in Figures 6 and 8 the temperature dependence of relaxation times shows non-Arrhenius-type activation behavior in agreement to the results on various

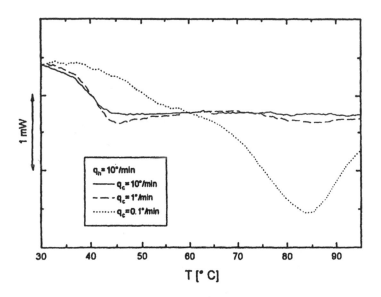

Figure 7   DSC thermograms of PC25 recorded during heating with a heating rate $q_h = 10\,K/$min. Previously, the samples have been cooled down from a temperature 40 K above T_g with cooling rates q_c of 10 K/min (solid line), 1 K/min (dashed line) and 0.1 K/min (dotted line).

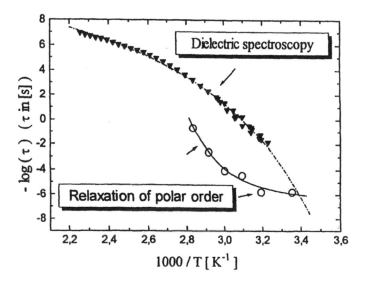

Figure 8   Relaxation times determined from dielectric spectroscopy and polar relaxation experiments for a sample of PC25. The cooling rate was 0.1 K/min in the poling procedure. The dashed–dotted line shows the fit of the dielectric relaxation times to the WLF equation.

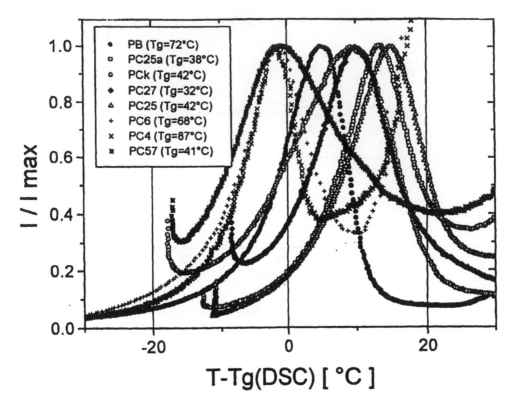

Figure 9 Comparision of TSDC scans of all EO-polymers studied in this work. In all cases the samples were poled with 40 V/μm above T_g and cooled down to room temperature with a cooling rate of 1 K/min. Glass temperatures T_g were obtained from the analysis of the DSC thermograms, which had been recorded with a cooling and heating rate of 10 K/min.

guest–host, side-chain and main chain polymers. Non-Arrhenius-type behavior can be also predicted based on the AGV equation (2), with T_f assumed to be temperature dependent.

Temperature dependent decoupling of dopant motion from the behavior of the host polymer has been argued recently in order to explain the Arrhenius type time–temperature characteristics for a 1 wt% guest–host system of 4-(dimethylamino)-4′-nitrostilbene (DANS) in poly(ethyl methacrylate) below the glass transition temperature [6]. An Arrhenius-type temperature dependence of relaxation has been also observed by studying the re-orientation dynamics of disperse red 1 (DR1) chromophores doped at 2 wt% in polystyrene [2]. Arrhenius-type activation behavior can be also explained by the AGV equation if the fictive temperature T_f remains constant in the time and temperature range studied. The change of T_f with time is roughly given by

$$\frac{dT_f}{dt} = \frac{T - T_f}{\tau}.$$

(4)

The structural relaxation time τ itself is a function of the actual temperature and of T_f. A rather constant T_f will be observed either if the sample has been quenched rather quickly from a temperature T_1 (above T_g) to the actual temperature or if the sample is kept at rather low temperatures implying a large value for τ. The temperature dependence of relaxation times indeed approaches Arrhenius-type activation ,at temperatures well below T_g. The type of the apparent activation behavior can thus be influenced by the specific thermal history of the sample.

Arrhenius-type activation of polar relaxation has also been observed for rigid-rod like polymers with head-embedded EO-chromophores [16,40]. In this case the motion of the chromophore can be decoupled from the dynamics in the side chain region due to the microphase separation into main chain layers and side chain layers. It was concluded that the relaxation is dominated by a rotation of the chromophore around the stiff polymer backbone. Such locally activated chromophore relaxation can be proposed to be a common phenomenon in layered structures of rigid-rod main chain polymers with chromophores rigidly linked to the main chain.

CONCLUSION

The data presented in this paper demonstrate that the relaxation kinetics of poled polymers in the glassy state are strongly connected on the thermal history. The degree of aging as determined by DSC, isothermal relaxation or TSDC experiments depended on the particular chemical structure. A correlation or comparison between data obtained by different methods such as dielectric spectroscopy or isothermal decay experiments is thus only meaningful, if the samples have undergone the comparable thermal cycles or if the glassy state has been sufficiently character-ized. We demonstrate that the type of the apparent activation behavior can be influenced by the specific thermal history of the sample.

ACKNOWLEDGMENTS

The authors like to thank Prof. G. Wegner, Dr. S. Schrader and Dr. R. Richert for many fruitful discussions. This work was in part funded by the German Ministry of research and technology under the project number 03 M 4046.

REFERENCES

[1]   M. Stähelin, C. A. Walsh, D. M. Burland, R. D. Miller, R. J. Twieg and W. Volksen, *J. Appl. Phys.* **73** (1993) 8471.
[2]   A. Dhinojwala, G. K. Wong and J. M. Torkelson, *J. Chem. Phys.* **100** (1994) 6046.
[3]   H.-T. Man and H. N. Yoon, *Adv. Mat.* **4** (1992) 159.
[4]   A. Dhinojwala, G. K. Wong and J. M. Torkelson, *Macromolecules* **26** (1993) 5943.
[5]   W. Köhler, D. R. Robello, P. T. Dao, C. S. Willand and D. J. Williams, *Mol. Cryst. Liq. Cryst. Sci. Technol.-Sec. B* **3** (1992) 83.
[6]   S. Schüssler, R. Richert and H. Bässler, *Macromolecules* **28** (1995) 2429.

[7]   C. Weder, P. Neuenschwander, U. W. Suter, P. Pretre, P. Kaatz and P. Günter, *Macromolecules* **28** (1995) 2377.

[8]   K. Zimmermann, F. Ghebremichael and M. G. Kuzyk, *J. Appl. Phys.* **75** (1994) 1267.

[9]   H. Guan, C. H. Wang and S. H. Gu, *J. Chem. Phys.* **100** (1994) 8454.

[10] L.-Y. Liu, D. Ramkrishna and H. S. Lackritz, *Macromolecules* **27** (1994) 5987.

[11] M. A. Firestone, M. A. Ratner, T. J. Marks, W. Lin and G. K. Wong, *Macromolecules* **28** (1995) 2260.

[12] G. A. Lindsay, R. A. Henry, J. M. Hoover, A. Knoesen and M. A. Mortazavi, *Macromolecules* **25** (1992) 4888.

[13] M. A. Firestone, M. A. Ratner and T. J. Marks, *Macromolecules* **28** (1995) 6296.

[14] D. M. Burland, R. D. Miller and C. A. Walsh, *Chem. Rev.* **94** (1994) 31.

[15] H. W. Spiess, *Chem. Rev.* **91** (1991) 1321.

[16] C. Heldmann, D. Neher, H.-J. Winkelhahn and G. Wegner, *Macromolecules* **29** (1996) 4697.

[17] H.-J. Winkelhahn, T. K. Servay, L. Kalvoda, M. Schulze, D. Neher and G. Wegner, *Ber. Bunsenges. Phys. Chem.* **97** (1993) 1287.

[18] M. Pfaadt, Ph.D. thesis, Johannes-Gutenberg Universität, Mainz (1995).

[19] T. K. Servay, H.-J. Winkelhahn, M. Schulze, C. Böffel, D. Neher, G. Wegner and L. Kalvoda, *Ber. Bunsenges. Phys. Chem.* **97** (1993) 1272.

[20] Th. K. Servay, Ph.D. thesis, Johannes-Gutenberg Universität, Mainz (1994).

[21] D. Riedel, Diploma thesis, Johannes-Gutenberg Universität, Mainz (1994).

[22] H. Rengel, Ph.D. thesis, Johannes-Gutenberg Universität, Mainz (1995).

[23] C.-S. Kang, H.-J. Winkelhahn, M. Schulze, D. Neher and G. Wegner, *Chem. Mat.* **6** (1994) 2159.

[24] H.-J. Winkelhahn, H. H. Winter and D. Neher, *Appl. Phys. Lett.* **64** (1994) 1347.

[25] H.-J. Winkelhahn, S. Schrader, D. Neher and G. Wegner, *Macromolecules* **28** (1995) 2882.

[26] J.D. Ferry, "*Viscoeleastic Properties of Polymers*", John Wiley & Sons, New York, (1980).

[27] H.-J. Winkelhahn, T. K. Servay and D. Neher, *Ber. Bunsenges. Phys. Chem.* **100** (1996) 123.

[28] O. S. Narayanaswamy, *J. Am. Ceram. Soc.* **54** (1971) 491.

[29] A. J. Tool, *J. Am. Ceram. Soc.* **29** (1946) 240.

[30] G. W. Scherer, "*Relaxation in Glasses and Composits*", John Wiley & Sons, New York, (1985).

[31] I. M. Hodge and A. R. Berens, *Macromolecules* **15** (1882) 762.

[32] A. J. Kovacs, J. J. Aklonis, J. M. Hutchinson and A. R. Ramos, *J. Poly. Sci. Polym. Phys.* **17** (1979) 1097.

[33] G. W. Scherer, *J. Am. Ceram. Soc.* **67** (1984) 504.

[34] G. Adams and J. Gibbs, *J. Chem. Phys.* **43** (1965) 139.

[35] A. K. Doolittle, *J. Appl. Phys.* **22** (1951) 1471.

[36] A. K. Doolittle, *J. Appl. Phys.* **23** (1952) 236.

[37] H.-J. Winkelhahn and D. Neher, *Proc. ICONO'2 Conference*, to appear in *Nonlinear Optics.*

[38] P. Kaatz, P. Petre, U. Meier, U. Stalder, C. Bosshard, P. Günter, B. Zysset, M. Stähelin, M. Ahlheim and F. Lehr. *Macromolecules* **29** (1996) 1666.

[39] H.-J. Winkelhahn, Ph.D. thesis, Johannes-Gutenberg University, Mainz (1995).

[40] C.-S. Kang, C. Heldmann, H.-J. Winkelhahn, M. Schulze, D. Neher, G. Wegner, R. Wortmann, C. Glania and P. Krämer, *Macromolecules* **27** (1994) 6156.

10. DESIGN AND FABRICATION OF ELECTRO-OPTIC POLYMER WAVEGUIDE DEVICES

JANG-JOO KIM and WOL-YON HWANG

Electronics and Telecommunications Research Institute,
P.O. Box 106, Yusong, Taejon 305-600, Korea

ABSTRACT

A couple of general guidelines for the design of electro-optic poled polymer waveguide devices are suggested. Firstly, the refractive index of one buffer layer needs to be close to the guiding layer to control the mode size and that of the other buffer layer to be much smaller than the guiding layer to lower the modulation voltage without the sacrifice of the mode size. In the asymmetric waveguide structure, the mode size is found to be solely determined by one buffer layer which has closer refractive index to the guiding layer. Secondly, the guiding layer needs to be fabricated as thick as possible within the range of the single mode operation for a given material system.

INTRODUCTION

Electro-optic (EO) poled polymers have attracted large attention during the last decade. Organic EO materials offer significant advantages over inorganic materials such as $LiNbO_3$ or compound semiconductors. They include large EO coefficients, low dielectric constants with negligible dispersion from DC to optical frequencies and process simplicity [1,2]. Much research effort has been poured into molecular engineering to improve the material properties, such as linear and nonlinear optical properties and thermal stability, with great successes during the last several years [3–6]. Electro-optic modulators with tens of GHz bandwidth [7–9], multilevel modulators [10] and switches [11] have already been demonstrated. Another important advantage of EO poled polymers is the controllability of poling induced optic axis. The property has been utilized to fabricate various polarization controlling devices such as polarization filters and polarization converters [12–14].

With the large success of material research and the demonstration of the potential of electro-optic poled polymer waveguide devices, it is an appropriate time to consider a detailed design of polymer waveguide devices in order to fully utilize the superior material properties [15]. In this paper, we will analyze the factors related to the design of the waveguide devices in detail and the experimental results based on the analysis will be shown.

WAVEGUIDE DESIGN

To analyze the issues involved in the design of polymer waveguide we consider a typical triple layer stacked electro-optic modulator whose cross sectional geometry is shown in Figure 1. The devices are considered to be poled with parallel plate electrodes so that the electro-optic coefficient along the direction normal to the substrate defined as Γ_{33} is the largest. The thickness and the refractive index of the guiding layer

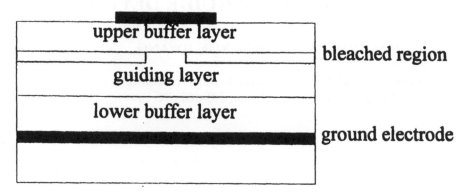

Figure 1   Cross-sectional geometry used for waveguide design.

are denoted as n_g and d_g, respectively. Those of the upper buffer and the lower buffer layers are n_u and d_u, and n_l and d_l, respectively. If one has a proper EO material for such a device, appropriate buffer materials have to be selected first and the thickness of each layer should be determined to fully utilize the EO material properties. For a demonstration purpose, we use the material of P2ANS/MMA $x/(100-x)$ as the guiding layer, which is a methacrylate backbone side chain copolymer of dimethylaminonitrostilbene (DANS) and methylmethacrylate (MMA), where $x/(100-x)$ is the molar ratio of DANS to MMA. The refractive index of the material with the composition of (50/50) is 1.635 at 1.3 μm before poling. The same kind of material with slightly lower chromophore composition was selected as the lower buffer layer. The refractive index of the material is 1.610 at 1.3 μm. The choice of the material as a buffer layer will become clear later.

Before any design of a channel waveguide, proper thicknesses of the guiding, lower and upper buffer layers should be determined by an analysis. The thickness range of the guiding layer is obtained from the single TM mode operation condition if the refractive indices of the buffer layers are given. This can be calculated from the simple dispersion relation. Figure 2 shows the dispersion relation calculated by the effective index method for various material systems at the 1.3 μm wavelength. The guiding layer thickness supporting the single mode operation increases as the asymmetry of the refractive index between the lower and the upper buffer layers increases. The dispersion relations of symmetric waveguides are included in the figure for comparison. The cut-off thickness for the single mode operation in the symmetric waveguides is thinner than the asymmetric cases even for $n_l = n_u = 1.617$.

The next step in the design is to find the thicknesses of the upper and lower buffer layers. These parameters affect two device operation parameters: modulation voltage and optical loss due to metal electrodes. The optical loss can be lowered by using thicker buffer layers. However, a thick buffer layer results in a high modulation voltage because it is inversely proportional to the total thickness of the waveguide.

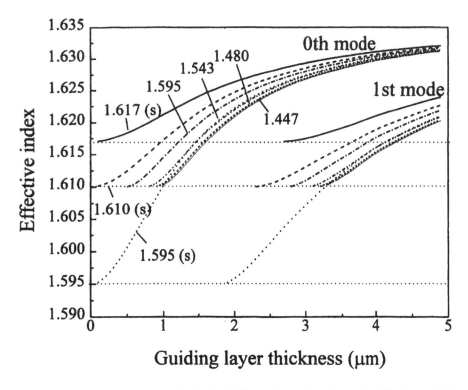

Figure 2 Dispersion relation of the device in Figure 1 as a function of guiding layer thickness for various refractive indices of one buffer layer. The refractive index of the other buffer layer is fixed to 1.610 except the numbers with (s) where the refractive indices of the buffer layers are the same.

We will try to find if there are any guidelines for the design of optimum device structure to overcome the trade off.

The electrode associated optical loss can be calculated by performing a complex mode analysis on the 5-layer waveguide structure of Figure 1 including the metal layers. The optical loss depends not only on the thickness of the buffer layers but also on the thickness of the guiding layer because the optical power confined in the guiding layer depends on its thickness. Figure 3 shows the total waveguide thickness to have a loss lower than a required limit, for instance 0.1 dB/cm, as a function of the guiding layer thickness within the single mode operation range. It is noteworthy that a thicker guiding layer actually results in a thinner overall structure required to satisfy the optical loss limit for a given material system. The figure also shows that a thinner waveguide is possible with lower refractive index buffer materials.

In order to analyze the modulation voltage of an electro-optic modulator, we have to consider the modulation efficiency. In general, buffer layer materials are not electro-optic so that an externally applied voltage to the waveguide induces the refractive index change of the guiding layer only. Since the wave propagating a

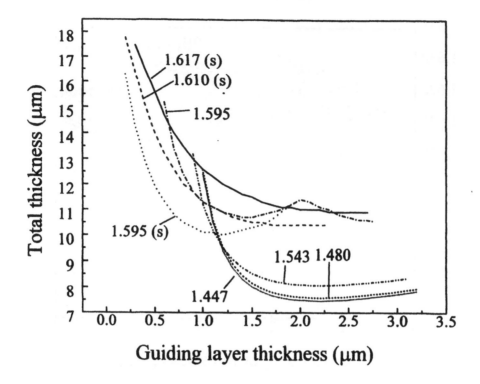

Figure 3 Total thickness of the waveguide in Figure 1 required to have electrode associated loss less than 0.1 dB/cm for different refractive indices of one buffer layer. The refractive index of the other buffer layer is fixed to 1.610 except the numbers with (s) where the refractive indices of both buffer layers are the same (symmetric waveguide).

waveguide feels the effective index change ΔN induced by the refractive index of the guiding layer Δn, it is important to take into account the modulation efficiency defined by $\Delta N/\Delta n$. It is calculated again from the dispersion relation against the guiding layer thickness and is shown in Figure 4. The modulation efficiency increases with the guiding layer thickness. This is consistent with expectation because the optical power confined in the guiding layer increases as the thickness of the guiding layer increases.

Modulation voltage of an EO modulator can be expressed by

$$V_\pi = \left(\frac{\lambda}{n^3 \Gamma_{33}}\right)\left(\frac{\Delta n}{\Delta N} \frac{d_{tot}}{L}\right) \tag{1}$$

where λ is the wavelength of the light, n the refractive index of the guiding layer material, d_{tot} the total thickness of the waveguide, and the L the length of the modulation electrode. The former part in the right-hand side of the equation is related to the material parameters and the later to the device design factors. For a

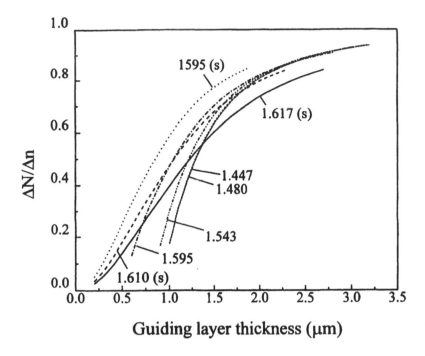

Figure 4 Modulation efficiency for different refractive indices of one buffer layer as a function of guiding layer thickness. The refractive index of the other buffer layer is fixed to 1.610 except the numbers with (s) where the refractive indices of both buffer layers are the same (symmetric waveguide).

modulator to have the lowest possible modulation voltage, one needs to maximize the device figure of merit (FOM), defined by $(\Delta N/\Delta n)/d_{tot}$, under a required loss limit due to the metal electrodes. The FOM can be obtained by combining Figures 3 and 4 and is in Figure 5. It clearly shows that the FOM increases as the thickness of the guiding layer increases. This fact suggests that the optimum device design can be attained with a thicker guiding layer within the single mode range for a given material system. It can also be found from the figure that the FOM increases by lowering the refractive index of one buffer layer while keeping the refractive index of the other buffer layer as 1.610 (or increasing the asymmetry between two buffer layers). The FOM increases more if both of the buffer materials have much lower index than the guiding layer. In such cases, however, the mode size shrinks significantly as the refractive index gets lower as will be discussed next.

The mode size of the guided light is an important factor in terms of the coupling loss with optical fiber whose mode size is approximately 8 μm at the wavelength of 1.3 μm. In Figure 6 are displayed the calculated mode sizes for various combinations of the refractive indices of the buffer layers. When the guiding layer is getting thicker, the mode size shrinks initially and passes through a minimum followed by a slow increase up to the boundary of the single mode region. The mode size we are

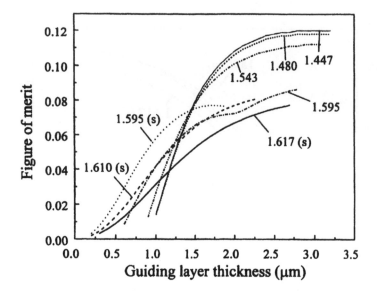

Figure 5 Figure of merit of an electro-optic modulator for different refractive indices of one buffer layer as a function of guiding layer thickness. The refractive index of the other buffer layer is fixed to 1.610 except the numbers with (s) where the refractive indices of both buffer layers are the same (symmetric waveguide).

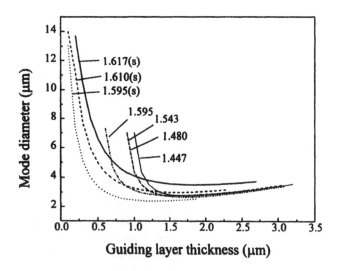

Figure 6 $1/e$ TM mode size in the vertical direction of symmetric (with (s)) and asymmetric waveguides (with (a)) for various combination of refractive indices of buffer layers as a function of guiding layer thickness. The refractive index of one buffer layer in the asymmetric waveguides is fixed to 1.610.

interested in is the one with a high FOM. The figure shows that exact mode matching with fiber is not possible even with the refractive index difference of 0.005 between the guiding and the buffer layers for a symmetric waveguide. However, the modulation voltage increases significantly with increasing the refractive index of the buffer layers as appeared in Figure 5. Decreasing the refractive index of the symmetric buffer layers from 1.617 to 1.595 results in rapid decrease of the mode size below 3 μm with only marginal improvement of modulation voltage. One way of maintaining a reasonable mode size and keeping the FOM high at the same time is to use an asymmetric structure. This is manifested by plotting the largest mode size and the maximum FOM for various combinations of buffer layers as shown in Figures 7 and 8, respectively.

Figure 7 shows that the mode size is solely determined by one buffer layer having the refractive index closer to the guiding layer whatever the refractive index of the other buffer layer is. The dips in Fig. 8 may be due to the surface plasmon. For instance, the mode size changes little as the refractive index of one buffer varies from 1.610 to 1.45 as long as that of the other buffer layer is maintained as 1.610. In other words, one can control the mode size of the waveguide by a proper selection of just one buffer layer. The other buffer material can be selected to decrease the modulation voltage.

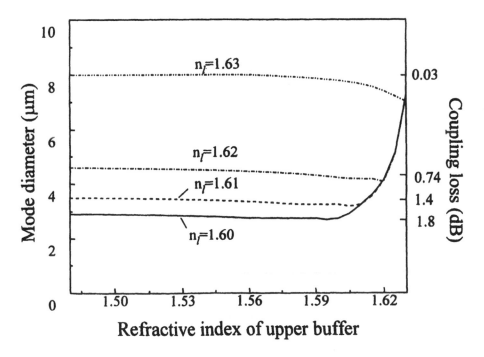

Figure 7   Variation of $1/e$ TM mode size as a function of refractive index of one buffer layer. The refractive indices of the other buffer layer are used parameters.

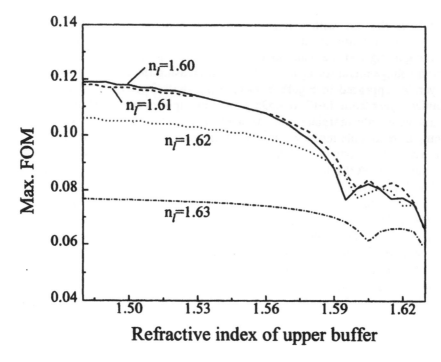

Figure 8 Maximum figure of merit for modulation voltage for various combinations of refractive indices of lower and upper buffer layers.

Selection of one buffer layer for a proper mode size should be done based on the loss budget for a device operation. The optical loss from fiber to fiber consists of propagation loss, coupling loss and electrode associated loss, and so forth. Among them the coupling loss is directly related to the mode size of a waveguide. The coupling loss is indicated in the right side ordinate of Figure 7. The coupling loss due to the mode mismatching is considered only for the vertical direction because the mode size of the guided light in the lateral direction can be controlled to have the same size as the optical fiber by adjusting the waveguide width and the photobleaching depth. The required minimum coupling loss due to the modal mismatch may vary for different applications. However it is hardly expected that it can be higher than 1–2 dB for each coupling. One should note that there are at least two couplings for a device. The refractive index of one buffer layer must be larger than 1.610 to have a coupling loss lower than 1.5 dB. This is the reason why the refractive index of one buffer layer is selected to be 1.610 in the previous calculations.

The other buffer layer can be selected based on the FOMs shown in Figure 8. The maximum FOM increases as the asymmetry of the refractive index between the two buffer layers increases. The results in Figures 7 and 8 indicate the buffer layer has to be selected to have much lower refractive index than the guiding layer. By using very low refractive index materials the thickness of the buffer layer satisfying the requirement of the electrode associated loss can be maintained very thin.

In summary, we can draw a couple of general guidelines for the design of EO polymer waveguides from the above analysis: Firstly one needs to fabricate the guiding layer as thick as possible within the region satisfying the single mode operation conditions. Secondly, the refractive index of one buffer layer needs to be close to that of the guiding layer to control the mode size, which in turn is related to coupling loss between the optical fiber and the waveguide. On the other hand the refractive index of the other buffer layer needs to be much smaller than the guiding layer to lower the modulation voltage.

The next step in the design is to form the channel waveguide. Photobleaching technique is considered for the design because of its simple process and good controllability. Even though photobleaching of a nonlinear optical polymer results in graded index profiles [16,17], the step index is assumed to be formed for simplicity. The waveguide structure in Figure 1 is used again for the design. The effective index method is also utilized in the lateral direction. Since the effective index in the bleached region changes as the bleaching proceeds, the effective index difference between the channel region and the bleached region changes with the bleaching depth. Following the nearly same procedure for the TE mode in the slab waveguide, the $1/e$ TM mode size in the lateral direction can be related to the bleaching depth as shown in Figure 9. In the calculation, the guiding layer thickness of $3.0\,\mu m$ and the waveguide width of $4\,\mu m$ are used. From the calculation

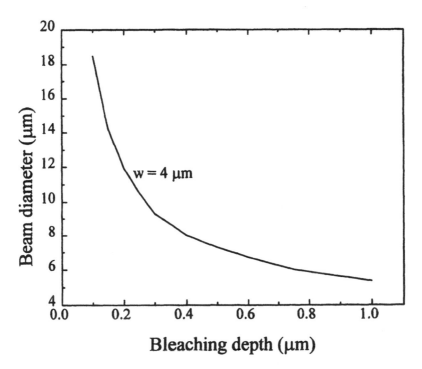

Figure 9 Change of $1/e$ TM mode size in the lateral direction as the bleaching proceeds for the waveguide width of $4\,\mu m$. The thickness of the guiding layer is $3\,\mu m$.

the bleaching depth can be determined to match the mode size with the optical fiber.

In the above analysis the birefringence induced by poling was not considered. It changes the numerical values appeared in the results. However it does not change the qualitative description extracted from the analysis.

FABRICATION OF ELECTRO-OPTIC MODULATOR

The material used for the guiding layer was described before. The material was supplied from Hoechst Celanese. An electro-optic coefficient of 30 pm/V is achievable with the material when the side group is sufficiently oriented by electric field poling. The refractive index of the material was measured by the m-line method at 1.3 µm, resulting in 1.635 before bleaching and 1.569 after complete bleaching. The same kind of copolymer with slightly different composition of (35/65) was used as the lower buffer material whose refractive index is 1.610. A hard optical epoxy (NOA 61) ($n = 1.543$) is selected as the upper buffer material. Gold coated Si wafers are used as the substrates. The lower buffer and guiding layers were successively coated on the substrates. The channel waveguides were formed by the irradiation of UV light through a waveguide mask using a mask aligner (Karl Suss MJB 3). After the channel waveguide formation the upper cladding layer was spin-coated. Thicknesses of the guiding, lower buffer and upper buffer layers were 3, 2.5 and 0.5 µm, respectively. Electrode poling was applied to align the chromophores after the completion of the waveguide structure. Finally the gold modulation electrode was formed on the waveguide. 1.3 µm light from a laser diode was coupled to the cleaved devices by the end fire coupling. A vidicon camera and a photodiode were used to analyze the mode pattern and the modulation characteristics.

RESULTS AND DISCUSSION

The modulation characteristics and the mode size of the EO modulator are shown in Figure 10. The device was poled with the poling field of 1.6 MV/cm. The modulation voltage of 6.7 V was obtained with the electrode length of 1.2 cm, which corresponds to the electro-optic coefficient of 23 pm/V. The extinction ratio was over 20 dB. The mode size of the guided light in the vertical direction was 5.3 µm that gives the coupling loss due to the modal mismatch less than 1 dB. When the upper buffer layer with a higher refractive index was used for the device, a higher modulation voltage with smaller mode size was resulted in. This fact demonstrates the importance of the proper selection of the buffer materials and the thicknesses.

When an EO material is poled, the refractive index increases along the poling field direction and decreases in the vertical direction. The resulting poled film becomes birefringent. Difficulties in the design of poled polymer waveguide devices are located in the fact that the refractive index of the EO polymer guiding layer changes with the poling field. The birefringence is proportional to the square of the poling field. Therefore the effect becomes significant especially for high poling fields. When we

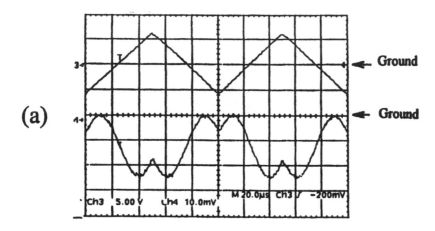

(a)

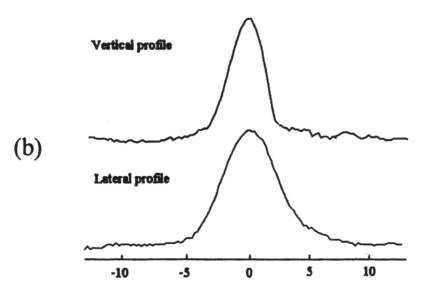

(b)

Figure 10 Modulation characteristics (a) and the mode sizes of the propagating light (b) of the electro-optic modulator fabricated following the guidelines suggested from the analysis. 20 kHz triangular shape electric signal with the peak to peak voltage of 12 V was applied to the device (ch3) and modulated optical output was recorded by a high speed photodetector (ch4). Extinction ratio is over 20 dB. /

performed the previous analysis including the poling induced birefringence for the device we fabricated, the device was in the multimode range in the vertical direction. Interestingly enough, however, only the zeroth mode was operational in the device. There are good reasons for the operation, which is reported in other paper [18].

Since the choice of cladding materials for the optimum utilization of an EO polymer material depends on the refractive index of the guiding layer, poling field for the device fabrication must be decided in prior. One way to avoid this complication is to use an EO material with slightly lower refractive index as a cladding material. The refractive index of the cladding material changes with poling as well as the guiding layer and the constraint can be relaxed much. The modulation efficiency can also be improved by using an electro-optic copolymer as a buffer layer compared to a linear buffer.

CONCLUSIONS

A couple of general guidelines for the design of electro-optic poled polymer waveguide devices are extracted. Firstly, the guiding layer needs to be fabricated as thick as possible within the range of the single mode operation for a given material system. Secondly, the refractive index of one buffer layer needs to be close to the guiding layer to control the mode size, and that of the other buffer layer to be much smaller than the guiding layer to lower the modulation voltage without sacrificing the mode size. In the asymmetric waveguide structure, the mode size is be solely determined by one buffer layer that has refractive index close to the guiding layer. Electro-optic modulators fabricated using the photobleaching technique support the guidelines.

ACKNOWLEDGMENT

This work was supported from the Ministry of Information and Communication of the Korea Government.

REFERENCES

1. J.-J. Kim and E.-H. Lee, *Mol. Cryst. Liq. Cryst.*, **227** (1993) 71.
2. R. Lytel, *SPIE Proceeding*, **1216** (1990) 30.
3. M. Eich, B. Reck, C. G. Wilson, D. Y. Yoon and G. C. Blorklund, *J. Appl. Phys.*, **66** (1989) 2559.
4. J. W. Wu, J. F. Valley, S. Ermer, E. S. Binkley, J. T. Kenney, G. F. Lipscomb and R. Lytel, *Appl. Phys. Lett.*, **58** (1991) 225.
5. C. V. Francis, K. M. White, G. T. Boyd, R. S. Moshrefzadeh, S. K. Mohapatra, M. D. Radcliffe, J. E. Trend and R. C. Williams, *Chem. Mater.*, **5** (1993) 403.
6. D. H. Hwang, J.-I. Lee, H. K. Shim, W.-Y. Hwang, J.-J. Kim and J.-I. Jin, *Macromolecules*, **24** (1994) 6000.
7. D. G. Girton, S. L. Kwiatkowski, G. F. Lipscomb and R. Lytel, *Appl. Phys. Lett.*, **58** (1991) 1730.
8. C. C. Teng, *Appl. Phys. Lett.*, **60** (1992) 1538.
9. W. Wang, D. Chen, H. R. Fetterman, Y. Shi, W. H. Steir and L. R. Dalton, *IEEE Photonics Technol. Lett.*, **7**(6) (1995) 638.
10. T. A. Tumobillo, Jr. and P. R. Ashley, *Appl. Phys. Lett.*, **62** (1993) 3068.

11. M. Hikita, Y. Shuto, M. Aano, R. Yoshimura, S. Tomura and H. Kozawaguchi, *Appl. Phys. Lett.*, **63** (1993) 1161.
12. M.-C. Oh, S.-Y. Shin, W.-Y. Hwang and J.-J. Kim, *Appl. Phys. Lett.*, **67** (1995) 1821.
13. W.-Y. Hwang, J.-J. Kim, M.-C. Oh and S.-Y. Shin, *IEEE J. Quantum Electron.*, **32** (1996) 1054.
14. J.-J. Kim, W.-Y. Hwang, M.-C. Oh and S.-Y. Shin, *SPIE Proceedings*, *397* (1995) 2527.
15. J.-J. Kim, W.-Y. Hwang and T. Zyung, *Mol. Cryst. Liq. Cryst.*, **267** (1995) 353.
16. J.-J. Kim, T. Zyung and W.-Y. Hwang, *Appl. Phys. Lett.*, **64** (1994) 3488.
17. T. Zyung, W.-Y. Hwang and J.-J. Kim, *Polymer Preprint*, **35** (1994) 285.
18. W.-Y. Hwang and J.-J. Kim, *Appl. Phys. Lett.*, **69** (1996) 1520 .

11. LARGE STABLE SECOND-ORDER COEFFICIENTS AND WAVEGUIDE DEVICE APPLICATION IN POLED SILICA FILM DOPED WITH AZO DYE

H. NAKAYAMA[a], O. SUGIHARA[b], and N. OKAMOTO[a]

[a] *The Graduate School of Electronic Science and Technology, Shizuoka University;* [b] *Faculty of Engineering, Shizuoka University, 3-5-1 Johoku, Hamamatsu 432, Japan*

ABSTRACT

Sol–gel processed silica films doped with organic azo dyes in a higher concentration of 40 wt%, are prepared. Large second-order nonlinear optical coefficients are obtained from azo dye doped silica film for the fundamental wavelength of 1064 nm, and the value is unvaried for 1000 h with no relaxation at 125°C. Moreover, ridge-type channel waveguide is fabricated, and Cerenkov-type phase-matched SHG is realized.

INTRODUCTION

Poled polymer films have the advantages of relatively large nonlinearity, ease of film formation, low cost for application to waveguide optical devices, and so on. Frequency doubling and electro-optic modulation with various poled polymer films have been studied. In particular, a guest–host system [1–5] is attractive because the nonlinear film can be fabricated with lower cost than for any other system, such as side-chain, cross-linked, and main-chain polymers, all of which are synthesized with more complicated reactions. However, in general, the chromophore concentration is low in guest–host polymer system, and the second-order nonlinearity decreases with time because of the orientation relaxation of the organic nonlinear molecules.

To obtain large and highly stable nonlinearity in guest–host system, the dye molecules having a large hyperpolarizability β should be mixed in high concentration and the chromophore orientation relaxation must be suppressed. One way to realize large and stable nonlinearity is to use sol–gel processed silica film doped with organic chromophores [6]. Sol–gel processing is an advantageous technique for preparing organic–inorganic films by a simple synthesis [7–11]. Organic chromophores soluble in the starting solution can be doped in virtually any ratio necessary to yield the desired properties, and electric field poling can be applied before the full densification of the glass film. Heat treatment with the poling field can lock the chromophores dispersed in numerous pores of the oxide matrix into any desired direction, and therefore large and temporally stable second-order nonlinearity is achieved.

In this study we prepare the silica films doped with organic azo dyes in higher concentration. Large second-harmonic generation (SHG) and its thermal stability are measured. Ridge-typed channel waveguide is fabricated with photolithography and etching technique, and Cerenkov-type phase-matched SHG is observed.

SAMPLE PREPARATION BY SOL–GEL METHOD

It was reported that the appropriate replacement of benzene-ring structures by thiazole rings in polar donor–acceptor molecules results in an increase in the β value [12]. Hence, if we choose an organic material with high β value, we may obtain new complex materials with high nonlinearity.

We prepared silica films doped with Disperse Red 1 dye (DR1: microscopic hyperpolarizability $\beta = 230 \times 10^{-30}$ esu, dipole moment $\mu = 7.99$ D) [13] and thiazole azo dye (TA: microscopic hyperpolarizability $\beta = 363 \times 10^{-30}$ esu, dipole moment $\mu = 8.13$ D) [14], whose structural formulas are shown in Figure 1, in a higher concentration with simple sol–gel processing.

The sol–gel processing of silica film doped with azo dyes are as follows [10]. Table 1 and Figure 2 show the composition of the starting solution and the flowchart for sample preparation, respectively.

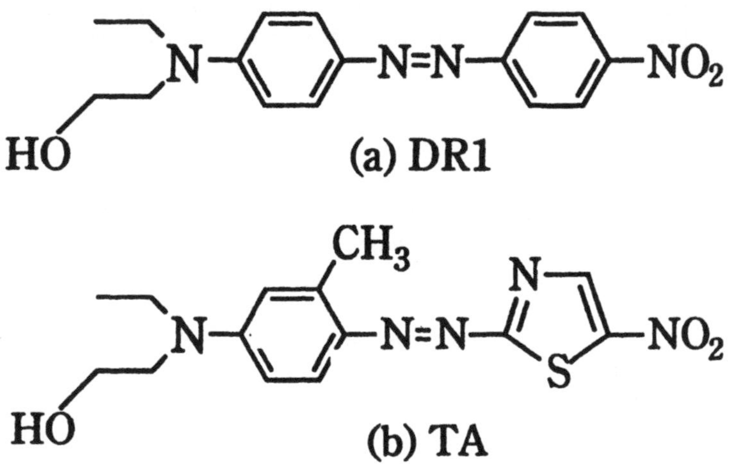

Figure 1 Structural formula of DR1 and TA.

Table 1 Composition of starting solution of sol–gel processing.

Material	Molar ratio
Azo Dye	$10 \sim 40$ wt%
$Si(OC_2H_5)_4$	1.0 mol
C_2H_5OH	1.0 mol
$(CH_3)_2NCHO$	1.7 mol
H_2O	3.8 mol
HCl	1.0 mol

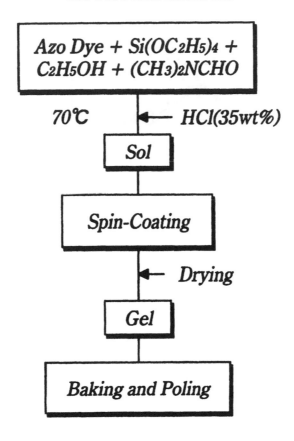

Figure 2 Flowchart for preparation of sol–gel processed silica film doped with azo dye.

Nonlinear organic materials (DR1, TA), tetraethoxysilane ($Si(OC_2H_5)_4$, TEOS), ethanol (EtOH), and N,N-dimethylformamide (($CH_3)_2NCHO$, DMF) were mixed and stirred for 30 min. Hydrochloric acid (35 wt%, HCl) was added to the solution, which was then stirred for about 2 min. The solution was spin-coated on a slide glass of 1 mm thickness. Good wetting ability of the surface of the slide glass is important for obtaining a homogeneous film. Therefore before spin-coating, the slide glass was soaked in an alkaline detergent solution, and then washed with pure water, and dried in blowing N_2 gas. After aging for 10 min at 70 °C, the temperature of the sample was gradually raised to 140 °C for an hour as shown in Figure 3. At the same time, corona poling was performed by applying a relatively high voltage of $+12$ kV to a needle electrode, then, to avoid spark discharge, the voltage was reduced to $+10$ kV by degrees. After applying a corona-discharge for an hour, the poled film was cooled to room temperature, maintaining the electric field, and then the field was removed. The space between the electrodes was 9 mm. On the assumption that all TEOS changes to silica, DR1 and TA could be doped in a high concentration up to 40 wt%.

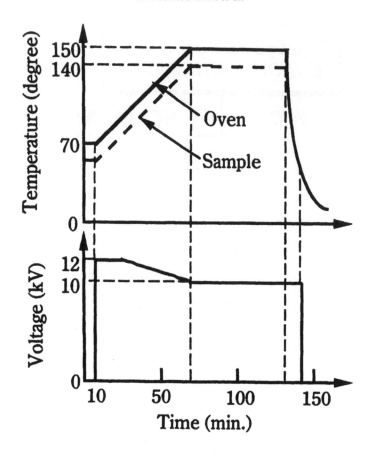

Figure 3 Patterns of temperature and applied voltage for baking and poling processes.

AZO DYE DOPED SILICA FILMS CHARACTERIZATION

The thickness of the prepared silica films were evaluated as 2.0 μm by means of the contact stylus method. The optical absorption spectra of the dye doped silica films were measured with a double beam spectrophotometer as shown in Figure 4. The solid lines indicate the absorption of unpoled silica film, and the dotted lines show the absorption of poled silica film, respectively. The peak absorbance wavelength and cut off wavelength of DR1/Silica and TA/Silica are $\lambda_{max} = 485$ nm, $\lambda_{cut} = 600$ nm (DR1) and $\lambda_{max} = 605$ nm, $\lambda_{cut} = 750$ nm (TA), respectively. It is notable that there is almost little absorption of TA doped silica film at blue light wavelength (around 450 nm), though there is strong absorption of DR1 doped silica film around it.

By the electric field poling process organic nonlinear chromophores were aligned in the poling direction and the cross-sectional absorption decreased as shown by dotted line. From Figure 4, the order parameters $\phi \, (= 1 - A_1/A_0)$, where A_1 is the absorbance of the poled sample and A_0 is the absorbance of the unpoled film, were

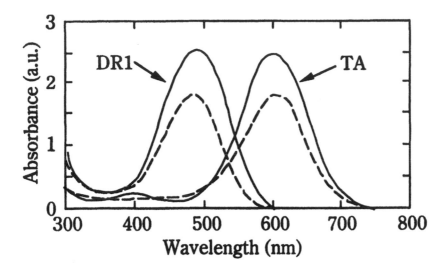

Figure 4 Optical absorption spectra of the dye-doped silica films before and after poling.

estimated as $\phi = 0.30$ (DR1) and $\phi = 0.21$ (TA), respectively at the each chromophore concentration of 40 wt%. The measured order parameter ϕ remained the same value after poling at room temperature and at 125°C for 1000 h.

The dispersion of the refractive index of the complex films are shown in Figure 5. The index was estimated from the Brewster angle measurements at 1064 nm, 820 nm, 800 nm, 633 nm, 594 nm, 544 nm. Sellmeir's equation which is indicated by solid lines was fitted as

$$1/(n^2 - 1) = C_0 + C_1/\lambda^2 \tag{1}$$

giving $C_0 = 0.772$, $C_1 = -9.22 \times 10^{-14}\,\mathrm{m}^2$ (DR1/Silica) and $C_0 = 0.693$, $C_1 = -6.42 \times 10^{-14}\,\mathrm{m}^2$ (TA/Silica), respectively. The indices at the fundamental and second harmonic wavelengths, 1064 nm and 532 nm for a Nd:YAG laser, were estimated as $n(\omega) = 1.568$, $n(2\omega) = 1.825$ (DR1/Silica) and $n(\omega) = 1.604$, $n(2\omega) = 1.773$ (TA/Silica), respectively.

SHG COEFFICIENTS AND STABILITY IN POLED SILICA FILMS

The second-order nonlinear optical coefficients were measured by the Maker fringe technique using a 1064 nm Q-switched (1 kHz) Nd:YAG laser with 90 ns pulse width and the peak input power 3.3 kW, and compared with the value of d_{11} of the quartz ($d_{11} = 0.5$ pm/V). Inasmuch as the harmonic wavelength (532 nm) is near the resonance peak of the azo dye doped silica film, Kleinman's equation in $C_{\infty v}$ symmetry is not satisfied. Therefore it is necessary to measure the each second-harmonic

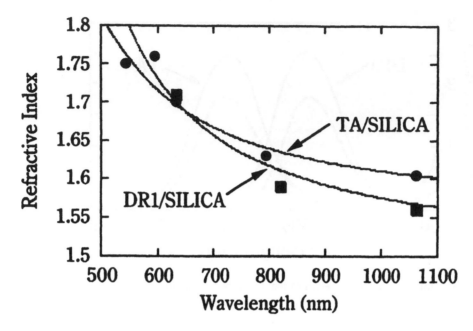

Figure 5   Refractive index of azo dye doped silica film versus wavelength.

generation tensor element (d_{31}, d_{15}, and d_{33}) independently. In this experiment we measured the generated harmonic intensity by changing the composition of the polarization angle of both the input fundamental wave and the output harmonic wave as follows [15]: (1) p-polarized fundamental wave and p-polarized harmonic wave (p–p); (2) s-polarized fundamental wave and p-polarized harmonic wave (s–p); and (3) input fundamental wave polarized differently by an angle φ from and an s-polarized harmonic wave (φ–s).

Figure 6 shows the SH intensity of the TA/Silica film measured by the measurement (1) as a function of incident angle. There is no fringe, because the silica film is so thin that the optical path length in the sample is shorter than the coherence length. The SH intensity increases with increasing incident angle and reaches a maximum at $\theta = 58°$.

The effective SHG coefficients of azo dye doped silica films measured by the measurement (1) as a function of dye concentration are shown in Figure 7. The effective d value of the film increased monotonously with the increase in azo dye concentration and reached maximum at 40 wt%. The peak effective d value were estimated as $d_{\text{eff}}^{\text{p-p}} = 40$ pm/V (DR1/Silica) and $d_{\text{eff}}^{\text{p-p}} = 44$ pm/V (TA/Silica), respectively. Micro crystal was observed when the concentration exceeded 40 wt%.

The largest second-order nonlinear coefficient d_{33} can be roughly estimated from the thermodynamic model [16–19] given by Eq. (2):

$$d_{33} = Nf^{2\omega}(f^{\omega})^2 \beta_{zzz} \langle \cos^3 \theta \rangle / 2, \tag{2}$$

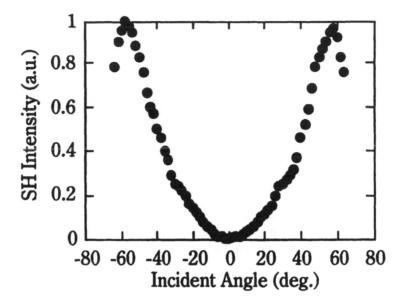

Figure 6   SH intensity of poled silica film versus incident angle.

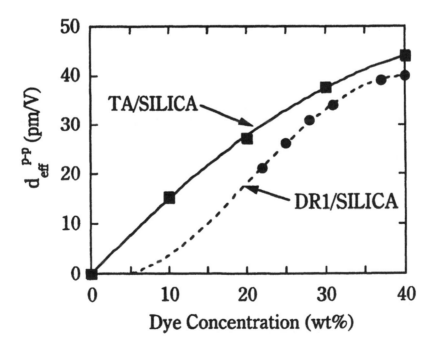

Figure 7   Effective SHG coefficient as a function of dye concentration.

where N is the doping density, f the local field factor, $\langle\ \rangle$ the expected value, β_{zzz} is the second-order molecular hyperpolarizability of the dye chromophore and θ is the angle between the dipole moment of the dye chromophore and the local electric field at the chromophore.

The poling method influences the macroscopic second-order polarization component through the amount of long range and long-term order that is imposed on the system. This is reflected in Eq. (2) by the term $\langle\cos^3\theta\rangle$. The orientation of the dipoles can be described by an order parameter. From Eq. (2) it follows that the statistical moments are

$$\langle\cos^n\theta\rangle = L_n(-\Delta W/kT),\tag{3}$$

where L_n is the nth-order Langevin function [20]. Consistent with the assumption that only dipole orientation effects are present, the first nontrivial axial order of the dipoles can be described by an order parameter [21]

$$\Phi = (3\langle\cos^2\theta\rangle - 1)/2.\tag{4}$$

From the measured order parameter Φ which was measured by spectroscopic absorption measurements, a chromophore alignment factor $\langle\cos^3\theta\rangle$ of DR1/Silica and TA/Silica was determined to be 0.31 and 0.35, respectively. From Eq. (2), nonlinear optical coefficient d_{33} of 138 pm/V and 168 pm/V were calculated, respectively. SHG from corona-poled silica films were measured with the Maker fringe technique using Nd:YAG laser in comparison with the value of d_{11} of the quartz. From the measurement of (1)–(3), the maximum second-order nonlinear optical coefficients were evaluated as $d_{31} = 26$ pm/V, $d_{15} = 27$ pm/V, $d_{33} = 120$ pm/V (DR1/Silica) and $d_{31} = 26$ pm/V, $d_{15} = 17$ pm/V, $d_{33} = 154$ pm/V (TA/Silica), respectively at the each chromophore concentration of 40 wt%. These observed values of d_{33} reasonably close to the calculated values. These values of d_{33} are one of the largest nonlinearity in guest–host system. As a comparison, DR1 and TA doped in the PMMA films were prepared and SHG coefficient was measured. Only 12.5 wt% of chromophores could be doped in PMMA matrix and SHG coefficients were measured as $d_{33} = 37$ pm/V (DR1/PMMA) and $d_{33} = 55$ pm/V (TA/PMMA), respectively.

Table 2 shows the values of SHG coefficients, hyperpolarizability and dipole moment of each samples.

Table 2 The value of SHG coefficient d_{33} and hyperpolarizability β and dipole moment μ (at 532 nm, 40 wt%).

Sample	d_{33} (exp.)	d_{33} (cal.)	$\beta\,(\times\,10^{-30}\,\text{esu})$	μ (D)
DR1/Silica	120 pm/V	138 pm/V	230	7.99
TA/Silica	154 pm/V	168 pm/V	363	8.13

It is important to investigate the thermal stability of the nonlinearity of azo dye doped silica film at 100–125°C, the temperature at which waveguide devices will be processed. Figure 8 shows the time dependent decay of the d_{33} value of the DR1 doped silica film measured at 125°C. We have confirmed that the d value was stable for at least 2000 h at room temperature without initial or long-term relaxation. Moreover, it is clear that the d value is stable up to 125°C without any relaxation for at least 1000 h. Similar property of stable nonlinearity was observed from TA doped silica film at 125°C.

Thus, it was found that both large nonlinearity and good thermal stability were obtained from dye doped poled silica film through simple sol–gel processing. The reasons may be as follows: In the process of sample preparation dye molecules hardly sublime and can remain in the film, thus a high net concentration of 40 wt% was obtained. Moreover, silica network is established during poling, so effective dipole alignment is obtained before the full densification of glass. These findings indicate larger nonlinear molecules aligned in the direction of poling field and are trapped in the pores. It is thought that for a combination of organic dye and silica the pore shrinkage by the high baking temperature stabilizes the nonlinearity in particular. It is also thought that hydrogen bonding can lead to an increase in stability by establishing a cross-linked network [22].

To investigate the resonance enhancement of the nonlinearity the wavelength dependent SHG coefficient of TA doped thin film was measured using an optical parametric oscillator (OPO) which is shown in Figure 9. In this measurement, TA

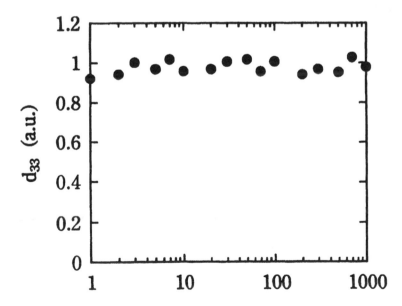

Figure 8 Thermal stability of nonlinear optical coefficient at 125°C.

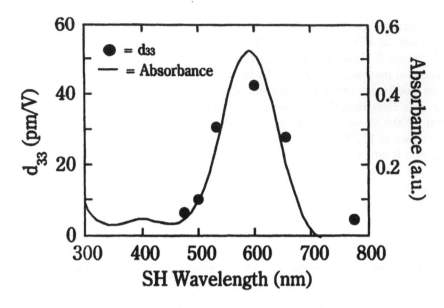

Figure 9   The d_{33} value versus the SH wavelength measured by OPO.

doped PMMA film (10 wt%, 3.1 μm) was used to facilitate the OPO measurement. The absorption spectrum of the film is also illustrated in the figure. The d_{33} value shows the maximum at 600 nm, and the peak value is about ten times larger than that in the nonresonant region. From the figure, the wavelength dispersion relation was found to strongly reflect the absorption spectrum of the film. The harmonic frequency with a S_0–S_2 transition of TA doped thin film leads to strong SH-enhancement. A similar dispersion relation will be obtained in azo dye doped silica film.

CHANNEL WAVEGUIDE FABRICATION

It is necessary to carry out minute processing in order to apply the poled complex silica film to the waveguide-type devices. Therefore, in this study, we attempted to fabricate a channel waveguide of DR1 doped silica film and to observe Cerenkov-type phase-matched SHG. There are some ways to fabricate a channel waveguide, such as ordinal method to use photolithography and etching techniques and UV-photobleaching method. This time we tried the former approach, and the latter approach will be reported elsewhere [23]. We fabricated a single-mode ridge-type channel waveguide. The fabrication process of the waveguide is as follows. Negative-type photoresist (OMR-83) was spin-coated onto the poled complex film deposited on the slide-glass substrate. After setting a photomask on the sample, UV-irradiation by a mercury lamp and development with rinse were performed to form a photoresist

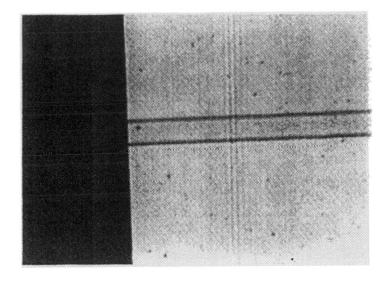

Figure 10   Top view of microscope photograph of fabricated channel waveguide.

channel. Lastly the complex film was sputter etched in Ar and O_2 gas to transcribe the photoresist pattern. The microscopic photograph of fabricated channel waveguide was shown in Figure 10. Since the poled complex silica film showed excellent resistance against most solvents such as methanol, MIBK and acetone, the spin-coating and development were very easy to be carried out without peeling off of the film nor flowing out of the dye molecules. In addition, the film has excellent thermal stability as shown in Figure 8, therefore no decrease of the second-order nonlinearity was maintained during etching process. The fabricated channel width and channel depth were estimated as $w = 8.0\,\mu m$ and $t = 0.52\,\mu m$. Using the perturbation method and mode dispersion curves, the channel waveguide was found to satisfy the condition of single-mode propagation for extraordinal wave of Nd:YLF laser (1047 nm). Next we tried to realize Cerenkov-type phase-matched SHG using the fabricated channel waveguide. In this experiment of Cerenkov-type SHG, a Nd:YLF laser with 4.7 kW peak power and 16 ns pulse width at a repetition frequency of 1 kHz was used as a fundamental wave. The incident beam was coupled into the waveguide with a prism (FD11) passing through a focusing lens. The propagation length was set as 5 mm. The Cerenkov-type SH wave (524 nm) was detected by a camera through a fundamental wave cut filter. The photograph of the far-field pattern of SH wave is in Figure 11. From the prism coupling-in angle of $\theta = 26.3°$, the effective index was estimated as $N_{eff} = 1.514$ which agrees well with theoretical calculation of mode dispersion curves.

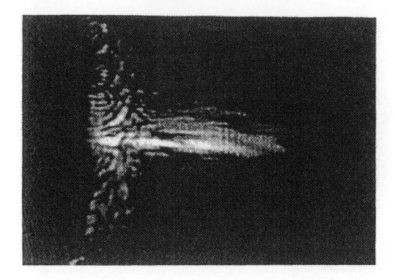

Figure 11 Photograph of Cerenkov-type phase matched SHG.

CONCLUSION

We have described the properties of the corona-poled silica film doped with the azo dye (DR1, TA) by the sol–gel method. The large nonlinear optical coefficients were evaluated as $d_{33} = 120\,\text{pm/V}$ (DR1/Silica) and $d_{33} = 154\,\text{pm/V}$ (TA/Silica) respectively at the each chromophore concentration of 40 wt%. The nonlinearity was stable for 1000 h at high temperature condition (125°C) without initial nor long-time relaxation. That is, these dye-doped silica films have both larger nonlinearity and greater thermal stability than the dye-doped polymer films and the sol–gel complex films reported so far. The reasons may be as follows. In the process of sample preparation azo dye molecules hardly sublime and can remain in the film, thus a high net concentration of 40 wt% was obtained. Moreover, the baking temperature of 150°C and the poling field of 12 kV are higher than in the other reports. These findings indicate larger nonlinearity in these films. The baking process with poling causes the pores in the oxide matrix to shrink and, hence, nonlinear molecules are aligned in the direction of poling field and are trapped in the pores. It is thought that for a combination of azo dye and silica the pore shrinkage and the high baking temperature stabilize the nonlinearity in particular.

However, drastical decay of nonlinearity was observed at more than 150°C due to decomposition of dyes, so if we choose appropriate guest material with high decomposition temperature, very stable second-order nonlinearity will be realized. Furthermore, we fabricated a single-mode ridge-type channel waveguide of DR1 doped silica film by using the photolithography and etching techniques as a practical approach to the application of this material, and realize Cerenkov-type phase matched SHG. We found that these azo dye doped silica films are well balanced material in its simplicity for film formation, nonlinearity, thermal stability, and mechanical and chemical strength sufficient for device fabrication. We are now trying the electro-optic modulation, using sol–gel processed silica film doped with organic chromophore.

ACKNOWLEDGEMENTS

The authors wish to thank H. Hayashi and M. Furukubo for their assistances of our study and their helpful discussions.

REFERENCES

[1]   K. D. Singer, J. E. Sohn and S. J. Lalama, *Appl. Phys. Lett.* **49** (1986) 248.

[2]   M. A. Mortazavi, A. Knoesen, S. T. Kowel, B. G. Higgins and A. Dienes, *J. Opt. Soc. Am. B* **6** (1989) 733.

[3]   H. L. Hampsch, J. M. Torkelson, S. J. Bethke and S. G. Grubb, *J. Appl. Phys.* **67** (1990) 1037.

[4]   O. Sugihara, T. Kinoshita, M. Okabe, S. Kunioka, Y. Nonaka and K. Sasaki, *Appl. Opt.* **30** (1991) 2957.

[5]   S. F. Hubbard, K. D. Singer, F. Li, S. Z. D. Cheng and F. W. Harris, *Appl. Phys. Lett.* **65** (1994) 265.

[6]   Y. Zhang, P. N. Prasad and R. Burzynski, *Chem. Mater.* **4** (1992) 851.

[7]   J. M. Boulton, J. Thompson, H. H. Fox, I. Gorodisher, G. Teowee, P. D. Calvert and D. R. Uhlmann, *Proc. Mater. Res. Soc. Symp.* **180** (1990) 987.

[8]   G. Puccetti, E. Toussaere, I. Ledoux and J. Zyss, *Polym. Prepr. Am. Chem. Soc. Div. Polym. Chem.* **32** (1991) 61.

[9]   Y. Zhang, P. N. Prasad and R. Burzynski, in *Chemical Processing of Advanced Materials*, L. L. Hench and J. K. West, eds. (Wiley, New York, 1992), p. 825.

[10] K. Izawa, N. Okamoto and O. Sugihara, *Jpn. J. Appl. Phys.* **32** (1993) 807.

[11] R. Burzynski and P. N. Prasad, in *Sol–Gel Optics: Processing and Applications*, L. C. Klein, ed. (Kluwer, Dordrecht, The Netherlands, 1994), p. 417.

[12] C. W. Dirk, H. E. Katz and M. L. Schilling, *Chem. Mater.* **2** (1990) 700.

[13] H. Hayashi, H. Nakayama, O. Sugihara and N. Okamoto, *Opt. Lett.* **20**(22) (1995) 2264.

[14] H. Nakayama, O. Sugihara, H. Fujimura, R. Matsushima and N. Okamoto, *Opt. Rev.* **2**(4)(1995) 236.

[15] L. M. Heyden, G. F. Saufer, F. R. Ore, P. L. Pasillas, J. M. Hoover, G. A. Lindsay and R. A. Henry, *J. Appl. Phys.* **68** (1990) 456.

[16] K. Yamaoka and E. Charney, *J. Am. Chem. Soc.* **94** (1972) 8963.

[17] J. Zyss and D. S. Chemla, (Academic, Orlando, Fla, 1987), Vol. 1, p. 3.

[18] D. J. Williams (Academic, Orlando, Fla, 1987),Vol. 1, p. 405.

[19] K. D. Singer, M. Kuzyk, and J. E. Sohn, *J. Opt. Soc. Am. B* **4** (1987) 968.

[20] P. Langevin, *Radium* **7** (1910) 250.

[21] A. Peterlin and H. A. Stuart, *Z. Phys.* **112** (1939) 129.

[22] C. Ye, T. J. Marks, J. Yang and G. K. Wong, *Macromolecules* **20** (1987) 2322.

[23] K. Ikegaya, I. Miyashita, O. Sugihara, N. Okamoto, C. Egami, *Nonlinear Optics* **15** (1996) 383.

12. NOVEL GUEST–HOST TYPE POLYMER FILMS FOR STABLE AND LARGE SECOND-ORDER NONLINEARITY AND WAVEGUIDE PROPERTY

M. OZAWA[a], M. NAKANISHI[a], H. NAKAYAMA[a], O. SUGIHARA[a], N. OKAMOTO[a], and K. HIROTA[b]

[a]Faculty of Engineering, Shizuoka University, 3-5-1 Johoku, Hamamatsu 432, Japan; [b]Research and Development Center, UNITIKA LTD, 23, Kozakura, Uji, Kyoto 611, Japan

ABSTRACT

New host polymers with high glass transition temperature doped with guest several dye in high concentration are prepared. Large and temporally stable second-harmonic generation (SHG) is observed, and optical waveguide properties are investigated.

INTRODUCTION

Recently nonlinear optical properties of poled polymeric materials have been investigated. Various types of poled polymers are prepared so far such as guest–host type, side-chain type, cross-linked type and so on. Especially poled guest–host polymer [1] has the advantage of low cost for film fabrication, but in general time dependent decay of its nonlinearity is fast, and second-order nonlinearity is small because of low chromophore concentration. For example, nonlinearity of host polymer (polymethylmethacrylate, PMMA) with relatively low glass transition temperature (109°C) doped with guest azo dye, Disperse Red 1 (DR1, $\beta = 230 \times 10^{-30}$ esu) drastically decreased to 40% of initial nonlinearity even after one week. On the other hand, poled side-chain type polymer has less relaxation and chromophore concentration is higher, but it becomes more expensive to synthesize such complicated polymers [2]. Recently, poled polymers with high glass transition temperature were reported [3,4], but its nonlinearity was small. In this study, new host polymer films with high glass transition temperature doped with several guest materials in high concentration are prepared. Large and temporally stable second-order nonlinearity is measured.

SAMPLE PREPARATION

Figure 1 shows chemical structures of new host polymers, U-100 and T-AP. The glass transition temperature of each polymer is 193°C (U-100), 233°C (T-AP), respectively. DR1, Chloroform ($CHCl_3$) and polymer (U-100 or T-AP) were mixed and stirred for about 30 min. The solution was spin-coated on a slide glass substrate with 1 mm thickness. Figure 2 shows the peak absorbance of the polymer film as a function of DR1 concentration. It indicates that Lambert–Beer's law is satisfied up

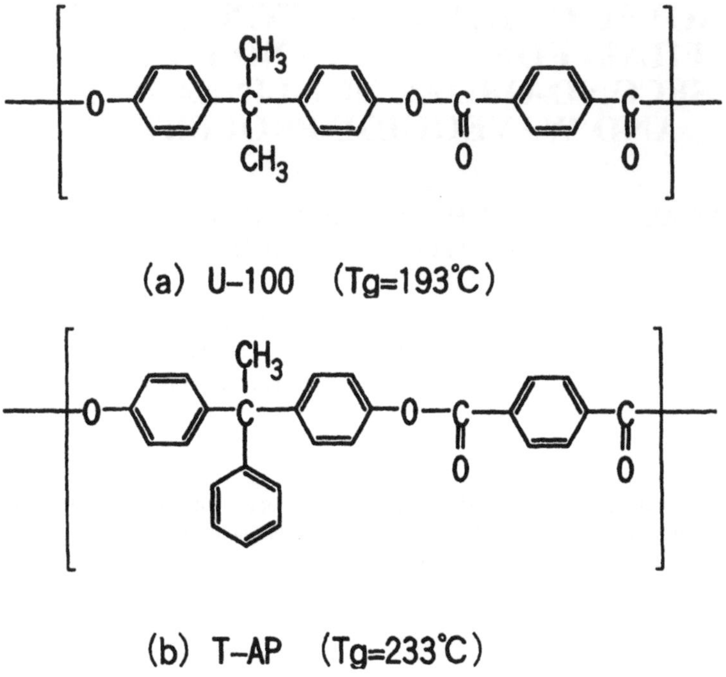

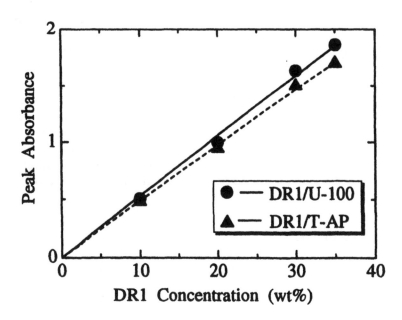

(a) U–100   (Tg=193℃)

(b) T–AP   (Tg=233℃)

Figure 1   The structure of polymer.

Figure 2   Peak absorbance versus DR1 concentration.

to 35 wt%, thus DR1 could be doped with a mono-molecular dispersion system up to 35 wt%. Its concentration is much higher in comparison with other guest–host system; for example DR1 doped PMMA system can have the maximum concentration of only 12.5 wt%. Corona-poling was performed by applying a high voltage to the thin films using a needle electrode.

SAMPLE NONLINEARITY

From absorption spectrum measurement corona-poled polymer films showed relatively large order parameter of $\Phi = 0.29$ for DR1/U-100 and $\Phi = 0.13$ for DR1/T-AP.

Second-harmonic generation (SHG) from corona-poled polymer films was measured with the Maker fringe technique using Nd:YAG laser (1064 nm) as a fundamental light source in comparison with the value of d_{11} of the quartz crystal ($d_{11} = 0.5$ pm/V). Note that these films have large absorption at the harmonic wavelength (532 nm), therefore the resonance-enhanced SHG coefficient is measured. We have found that the SHG coefficient increased linearly with increase in DR1 concentration up to 35 wt%. Maximum SHG coefficient is evaluated as high as $d_{33} = 100$ pm/V (DR1/U-100), $d_{33} = 65$ pm/V (DR1/T-AP) respectively as shown in Table 1. As a comparison SHG from corona-poled DR1/PMMA system was measured, and the nonlinearity was only 37 pm/V. In our measurement the smaller d_{33} and the smaller ratio d_{33}/d_{31} in DR1/T-AP system than in DR1/U-100 derive from lower poling temperature (140°C) than T_g which leads to negative condition of dipole alignment of DR1. This tendency was supported from each order parameter Φ.

Table 2 shows SHG coefficients of polymer film doped with other NLO dye doped film. We used 7-(4'-dimethylamino phenyl)-2-cyanopenta dienoic acid methyl (DACM), 5-(4-dimethylamino phenyl)-2-cyanopenta dienoic acid allylester (DACA), dimethylamino nitrostylbene (DANS) and 5-nitrothiazole-2-azo-[2'-methyl-4'-{N-ethyl-N-(2-hydroxyethyl)}-aminobenzene] (TA) as NLO dyes. Among them DACM/T-AP system showed relatively large nonlinearity large nonlinearity of 63 pm/V (resonance enhanced), and the ratio $d_{33}/d_{31} \sim 2.7$.

Table 1   The value of SHG coefficients d_{33} and d_{31}.

Polymer	DR1 concentration (wt%)	T_g (°C)	d_{33} (pm/V)	d_{31} (pm/V)
U-100	35	193	100	37
T-AP	35	233	65	29
PMMA	12.5	109	37	9

Table 2   The value of SHG coefficients of other NLO dye doped film.

Sample	Dye $\beta \times 10^{-30}$ esu	Dye concentration (wt%)	d_{33} (pm/V)	d_{31} (pm/V)
DACM/T-AP	>250	20	63	24
DACA/T-AP	243	20	31	10
DANS/T-AP	217	6	7.8	2.4
TA/U-100	363	<10	7.9	2.7

SAMPLE STABILITY

Figure 3 shows the time dependent decay of nonlinearity for each guest–host system. About 10 hours after poling, each nonlinearity decreased to 70%~90% of the original value and became stable for 1000 hours at room temperature thereafter. Nonlinearity of T-AP film doped with DACM (20 wt%) also decreased to 85% of original value and became stable for 1000 hours at room temperature thereafter. On the other hand nonlinearity of DR1 doped PMMA system (12.5 wt%) decayed to half even 20 hours after poling and it still continues to decay. The orientation relaxation of poled film becomes larger with increase in DR1 concentration, because the glass transition temperature of poled polymer doped with DR1 in a higher concentration is getting lower. Moreover, we tried to observe the orientation relaxation at higher temperature condition (80°C, 100°C). Figure 4 shows thermal decay of nonlinearity in DR1/T-AP system at high temperature.

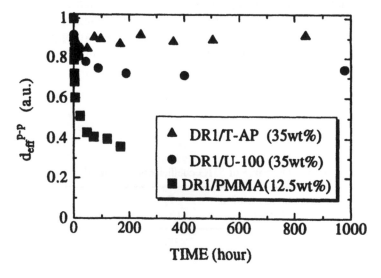

Figure 3   Time dependent decay at room temperature.

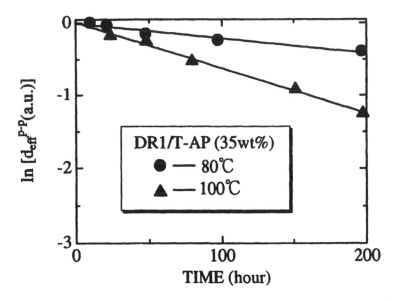

Figure 4 Time dependent decay.

From this figure the lifetime at room temperature was evaluated as long as $\tau = 2.3$ years (DR1/U-100) and $\tau = 7.1$ years (DR1/T-AP) respectively by arrhenius plot.

WAVEGUIDE PROPERTIES

It is important to investigate waveguide properties of DR1/T-AP guest–host system for the use of waveguide-type SHG and/or electro-optic devices. Figure 5 shows the propagation loss of DR1/T-AP (10 wt%) slab-type optical waveguide, which was measured with the scattering detection method using Nd:YLF (1047 nm) and He–Ne (632.8 nm) lasers. It shows that the propagation loss of 3.3 dB/cm at 632.8 nm and 2.8 dB/cm at 1047 nm was measured respectively. As a comparison the propagation loss of only T-AP waveguide was also measured, and 2.0 dB/cm was obtained.

Moreover, we have fabricated a single-mode channel waveguide by exposing UV light onto the clad region through a photo-mask; a useful method for fabricating the channel waveguide because the NLO chromophore is easily photobleached by UV light, and the developing and etching processes are unnecessary. The size of fabricated channel waveguide was 8.0 μm (width) and 0.7 μm (depth), and the step of the refractive indices between core- and clad-region is $\Delta n = 0.01$ (at 1047 nm).

Furthermore, we have observed Cerenkov-type frequency doubling using DR1/T-AP (10 wt%) waveguide. A fundamental wave of Nd:YAG laser (1064 nm) with 3 kW peak power and 90 ns pulse width at a repetition frequency of 1 kHz was coupled into a waveguide through a FD11 prism. The propagation length was set at 10 mm. The end-coupled harmonic wave (532 nm) was observed with the efficiency of

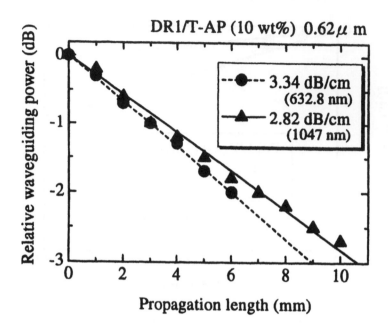

Figure 5 Propagation loss.

1.4×10^{-3} %/W. From the prism coupling-in angle of $\theta = 26.2°$, the effective index was estimated as $N_{eff} = 1.512$, which corresponds to the film thickness of $0.62\,\mu m$. This effective index satisfied well with the condition of Cerenkov-type radiation.

CONCLUSION

In conclusion corona-poled new polymer films doped with DR1 were prepared. The large SHG coefficient of $d_{33} = 100\,pm/V$ was obtained for the DR1/U-100 film and $d_{33} = 65\,pm/V$ for DR1/T-AP at the DR1 concentration of 35 wt%. Lifetime at room temperature is evaluated as long as $\tau = 2.3$ years (DR1/U-100), $\tau = 7.1$ years (DR1/T-AP) respectively by arrhenius plot. This system is superior to previously reported guest–host type poled polymers [1,4] in its high chromophore concentration and high glass transition temperature which leads to both large nonlinearity and long lifetime. Moreover, waveguide properties of propagation loss, fabrication of single-mode channel waveguide, and observation of Cerenkov-type frequency doubling were realized using poled DR1/T-AP system.

REFERENCES

[1] K. D. Singer, J. E. Sohn and S. J. Lamama, *Appl. Phys. Lett.* **49** (1986) 248.
[2] W. Sotoyama, S. Tatsuura and T. Yoshimura, *Appl. Phys. Lett.* **64** (1994) 2197.
[3] M. Stahelin, D. M. Burland, M. Ebert, R.D. Miller, B.A. Smith, R.J. Twieg, W. Volksen and C. A. Walsh, *Appl. Phys. Lett.* **61** (1992) 1626.
[4] Y. Matsuoka and A. Suzuki, *Kobunshi Ronbunshu.* **51** (1994) 442 (in Japanese).

13. EFFECT OF UV EXPOSURE ON POLED ORGANIC–INORGANIC FILMS FOR CHANNEL WAVEGUIDE FABRICATION

KOJI IKEGAYA, IKUYA MIYASHITA, OKIHIRO SUGIHARA,
NAOMICHI OKAMOTO, and CHIKARA EGAMI

Faculty of Engineering, Shizuoka University, 3-5-1 Johoku, Hamamatsu 432, Japan

ABSTRACT

A channel waveguide was fabricated by exposing UV light on sol–gel processed poled silica film doped with Disperse Red 1. We report the characteristics of the film and the mechanisms of UV light photo-bleaching for the waveguide fabrication.

INTRODUCTION

Poled organic–inorganic films [1–3] have the advantages of large nonlinearity, excellent thermal and chemical stability, good processability and so on. These films are useful for the application to the waveguide SHG and/or electro-optic devices. We have synthesized sol–gel processed poled silica films doped with organic azo dye [2,3], and large and stable second-order nonlinearity was obtained. However, there are very few reports of waveguide property using organic–inorganic films. In this paper we report the characteristics of the film and the mechanisms of UV light photo-bleaching for the channel waveguide fabrication. UV exposure for a waveguide preparation results in the decay of second-order nonlinear optical property and the decrease of refractive indices of the complex organic–inorganic film. This procedure is effective for step-index channel waveguide fabrication or $\chi^{(2)}$ grating formation. Moreover, we propose the easy way of making a traveling-wave type optical modulator using silica film doped with Disperse Red 1 (DR1).

SAMPLE CHARACTERISTIC

A channel waveguide using sol–gel processed poled silica film doped with DR1, DR1/Silica, was fabricated by UV exposure onto the film to induce the extraordinary index decrease of the clad region. Figure 1 shows the chemical structure of DR1. UV light with a beam power of $250\,W/cm^2$ from mercury lamp was exposed onto the complex organic–inorganic film with the dye concentration of 35 wt%. The complex film was synthesized with simple sol–gel processing. Orientational order to generate second-order nonlinearity was imported to this film by poling with a corona discharge during heat treatment. Large second-harmonic generation (SHG) in $d_{33} = 120\,pm/V$, which was thermally stable for 1000 hours at 125°C, was measured [4].

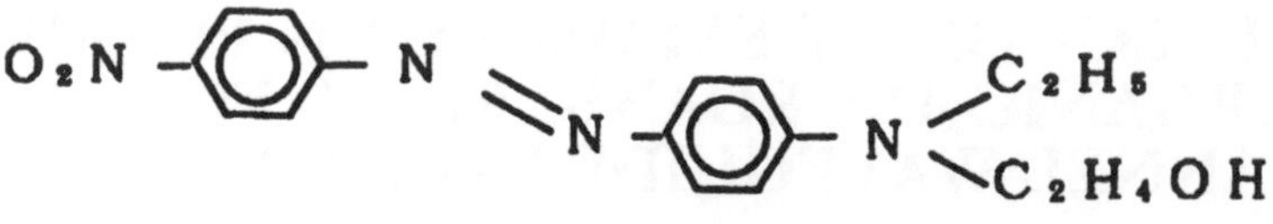

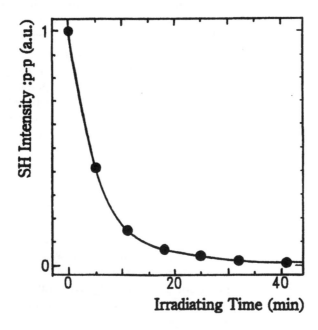

Figure 1   Structure of Disperse Red 1.

EXPERIMENTAL RESULTS

We measured the decay of second-order nonlinear optical property, the decrease of refractive indices and the change of the absorption spectrum of the complex film through UV exposure. By means of the Maker Fringe measurement SH intensity arising from second-order nonlinearity of the complex film exponentially decayed as applying the UV light and reduced by two orders of magnitude from the initial value for only 40 minutes exposure as shown in Figure 2. Moreover, it was found that from the measurement of the *m*-line method using Nd : YLF laser (1047 nm) the refractive indices of both TE mode and TM mode decreased in two processes. The rapid decrease in the first 4 hours was followed by the slow decrease in the next 16 hours (see Figure 3). The refractive index change of more than 0.04 was obtained by the UV exposure even in the first stage of the decay. Such a large change was enough for the use as the step index between core and clad regions of channel waveguide.

Figure 2   Second-order nonlinear optical property UV exposure time dependent decay of DR1/Silica.

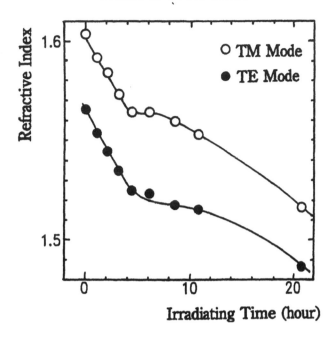

Figure 3   UV exposure time dependence of refractive index of DR1/Silica.

The absorption spectrum also changed in two processes. In the first stage the absorption peak at 500 nm decreased and the new absorption peak appeared at 370 nm (see Figure 4(a)). In the second stage these absorption peaks at both 500 and 370 nm decreased slowly (see Figure 4(b)). Figure 5 shows UV irradiation time dependence of absorption peaks at both 500 and 370 nm. As observed in the profile at 370 nm the absorption spectrum is found to change in two processes.

The mechanisms of both the decay of nonlinearity and the decrease of indices were investigated by the absorption spectrum, thin layer chromatography and H-NMR studies. In the first stage the cis–trans photo-isomerisation as well as photo-bleaching of DR1 molecules reduced the nonlinearity of the film. In this process the nonlinearity was not recovered by the alternate irradiation of visible and UV light. The UV exposure mainly photo-bleaches the $N = N$ chain of the DR1 chromophore, and the DR1 changes into two materials, one of which has p-Nitroaniline (p-NA)-like molecular structure with the absorption peak near 370 nm. The production of p-NA-like molecules results in the gradual increase of absorbance at 370 nm, as shown in Figure 5. In the following stage the generated p-NA-like material is sublimated in air and the absorption peak at 370 nm of the complex film decreased gradually. As a result the two processed refractive indices change of the complex film shown in Figure 3 arises from the production and the sublimation of p-NA-like molecules as stated above.

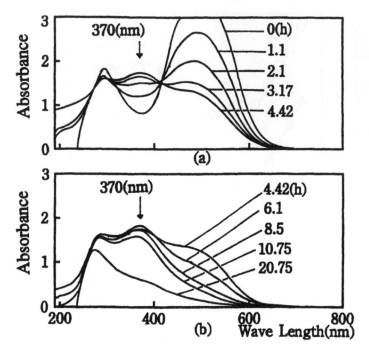

Figure 4 Change of absorption spectrum of DR1/Silica by UV exposure.

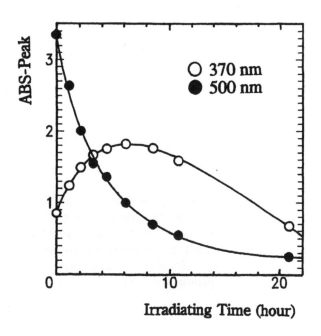

Figure 5 UV exposure time dependence of absorption peak.

FABRICATION OF CHANNEL WAVEGUIDE FOR TRAVELING-WAVE TYPE OPTICAL MODULATOR

We fabricated a single-mode channel waveguide by exposing UV light onto the silica film doped with DR1 through a photomask. The thickness of the channel waveguide was 1 μm and the width was 8 μm. Figure 6 shows the top view picture of microscopic observation of the channel waveguide.

This method of channel waveguide fabrication can be applied to the traveling-wave type optical modulator using the silica film doped with DR1. Figure 7 shows the cross-sectional structure of the traveling-wave type optical Mach–Zehnder modulator. First the DR1/Silica film is spin-coated on the substrate and corona-poling is applied to the whole region of film surface. Then thin transparent buffer layer is deposited onto the DR1/Silica film, and the electrode that has as same width as a channel waveguide is prepared. The electrode plays a role of photomask during UV exposure. The refractive index of the DR1/Silica film exposed to UV light is decreased and step index change is realized between clad and core regions resulting in the channel waveguide. In this method special process such as development and etching is unnecessary, and channel waveguide-based devices can be simply fabricated.

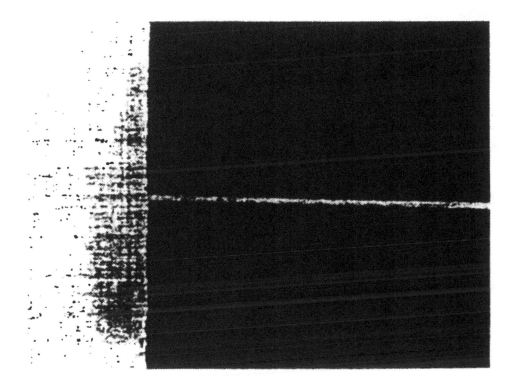

Figure 6 The top view picture of the channel waveguide.

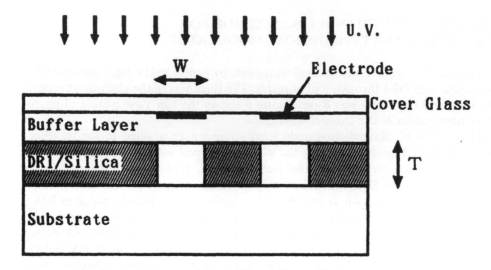

Figure 7   Structure of the traveling-wave type optical modulator.

CONCLUSION

We investigated the effect of UV exposure on sol–gel processed poled silica film with DR1 for the channel waveguide fabrication. This investigation made it clear that the change in absorbance and refractive indices of the film was caused by the production and the sublimation of the *p*-NA-like molecules by UV exposure. The UV exposure gives rise to the photo-bleaching of DR1 in the film, which is followed by the decay of second-order nonlinearity. The single-mode channel waveguide was fabricated, and we proposed the simple method of making the traveling-wave type optical modulator.

REFERENCES

[1]   Y. Zhang, P. N. Prasad and R. Burzynski, *Chem. Mater.*, **4**, 851 (1992).
[2]   K. Izawa *et al.*, *Jpn. J. Appl. Phys.*, **32**(2), 807–811 (1993).
[3]   H. Nakayama *et al.*, *Optical Review*, **2**(4), 236–238 (1995).
[4]   H. Hayashi *et al.*, *Optics Letters*, **20**(22), November 15, 1995.

14. HIGH PERFORMANCE CHROMOPHORES AND POLYMERS FOR ELECTRO-OPTIC APPLICATIONS

ALEX K.-Y. JEN[a,*], TIAN-AN CHEN[a], VARANASI P. RAO[a], YONGMING CAI[a], YUE-JIN LIU[a], and L.R. DALTON[b]

[a] ROI Technology, Optical Materials Division, 2000 Cornwall Rd., Monmouth Junction, NJ 08852; [b] Loker Hydrocarbon Institute, University of Southern California, Los Angeles, CA 90089-1661

INTRODUCTION

Organic polymeric electro-optic (E-O) materials have attracted significant attention because of their potential use as fast and efficient components of integrated photonic devices [1,2]. However, the practical application of these materials in optical devices is somewhat limited by the stringent material requirements imposed by the device design, fabrication processes and operating environments. Among the various material requirements, the most notable ones are large electro-optic coefficients (r_{33}), low optical loss and high thermal stability [3]. The design of poled polymeric materials with high electro-optic activity (r_{33}) involves the optimization of the number density of efficient (large $\beta\mu$) second order nonlinear optical (NLO) chromophores into a polymer matrix and the effective creation of poling-induced noncentrosymmetric structures. The factors that affect the material thermal stability are (a) the inherent thermal stability of the NLO chromophores, (b) the chemical stability of the NLO chromophores during polymer processing, and (c) the long-term dipolar alignment stability at high temperatures. Although considerable progress has been made in achieving these properties [4], organic polymeric materials suitable for practical E-O device applications are yet to be developed. This paper highlights some of our approaches in the optimization of molecular and material nonlinear optical and thermal properties.

THERMALLY AND CHEMICALLY STABLE NLO CHROMOPHORES DERIVED FROM THE 1,1-DICYANOVINYL ELECTRON ACCEPTOR

Our earlier research has clearly shown that large molecular nonlinearity ($\beta\mu$) can be achieved in donor-acceptor substituted stilbenes by replacing phenyl moieties with easily delocalizable thiophene moieties [5–8]. By combining thiophene conjugating units with tricyanovinyl electron-acceptors, highly efficient nonlinear optical chromophores have been developed [9]. For example, the measured nonlinear optical activity of compound 1 (Scheme 1) in dioxane solvent was 6200×10^{-48} esu at 1.907 µm (Under the same conditions, the measured activity of the standard

---

* Corresponding author. Current address: Department of Chemistry, Northeastern University, 360 Huntington Ave. Boston, MA 02115.

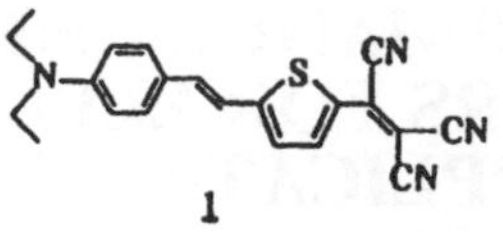

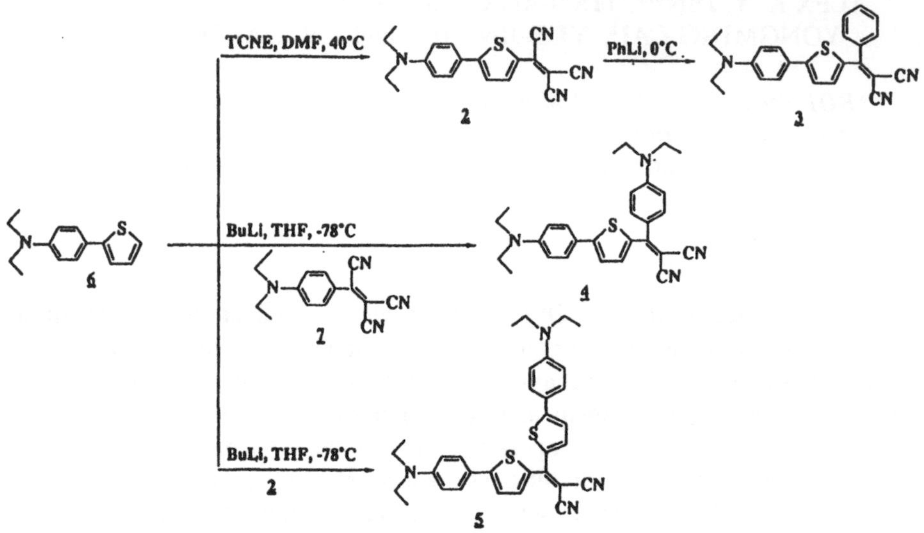

Scheme 1

dialkylamino-nitro stilbene, DANS, was 580×10^{-48} esu). The large molecular nonlinearity of **1** prompted us to evaluate its secondary properties and modify further to achieve the required tradeoffs.

Examination of the thermal stability of **1** using a sealed-pan DSC technique (heating rate 20°C/min) indicated its decomposition at 274°C. Clearly its thermal stability was not sufficient for practical device applications, and had to be improved to 300°C or above. It was anticipated that the presence of the isomerizable olefinic bond in **1** resulted in its relatively low thermal stability. Moreover, it is known that isomerization of the olefinic bond from its trans geometry to cis geometry causes localization of electron density and, thereby, such olefinic sites in chromophores are highly susceptible to electrophilic attack by singlet oxygen. As a first modification, compound **2**, an analog of **1** without the olefinic bridge, was developed [10]. This led to lowering of the $\beta\mu$ value (2700×10^{-48} esu) with respect to **1**, but enhanced the thermal stability to 300°C as anticipated.

Having achieved an excellent tradeoff between thermal stability and molecular nonlinearity with compound **2**, we attempted to incorporate **2** into high temperature polyimides (Hitachi PQ-2200 and Ultradel 4212) as solid solutions [11]. Such studies revealed several other limiting factors associated with compounds containing tricyanovinyl groups. First, concentration of the guest in host polyimides could not be optimized beyond 10–15% because of its poor solubility. Secondly, the tricyanovinyl group was found to be very sensitive to the polyamic acid curing conditions. Such a decomposition process lowers the effective concentration of the chromophore in fully cured polyimides. In another control experiment, **2** was heated in some of the processing solvents of polyamic acids, such as N-methyl pyrrolidine (NMP) and N,N-dimethylacetamide (DMAc). To our surprise, ready decoloration of the chromophore occurred at the boiling temperatures of the solvents NMP and DMAc. Analysis, indicated that the tricyanovinyl group was highly sensitive to these conditions, although the mechanism of the decomposition is not apparent.

To overcome these deficiencies associated with the tricyanovinyl group, **2** was further modified by replacing the most reactive CN in the tricyanovinyl group with aryl groups. It was anticipated that the bulky aryl groups would enhance the overall thermal stability and protect the resultant dicyanovinyl group from any nucleophilic substitution reactions that might take place during the processing/curing of polyamic acids. This modification would result in lowering the electron-deficiency of the acceptor-end and thus, the overall molecular nonlinearity. To test these concepts, compounds **3–5** were developed. The compound **3** was prepared by treating the tricyanovinyl derivative **2** with PhLi. A synthetic sequence involving the generation of the lithium derivative of **6** and its subsequent reaction with the tricyanovinyl derivative **7**, was used to make **4**. The compound **5** was made by reacting the lithium derivative of **6** with compound **2**. All the new compounds were fully characterized using standard spectroscopic and analytical techniques.

Analysis of the data revealed that the replacement of the cyano group on the vinylcarbon-2 of the tricyanovinyl acceptor of **2** with a phenyl moiety resulted in a significant blue shifted charge-transfer absorption and a lower $\beta\mu$ value. However, the introduction of an electron-donor such as dialkylamino group on the benzene ring of **3** (i.e., compound **4**) caused a significant increase in $\beta\mu$ (Table 1). Further increase in molecular nonlinearity was obtained by extending the conjugation in **4**

Table 1   The first molecular electronic hyperpolarizability and thermal stability data obtained for nonlinear optical compounds 1–5.

Compound[a]	λ_{max}(nm)	$\beta\mu \times 10^{48}$(esu)[b]	T_d(°C)[c]
1	653	6200	274
2	607	2700	296
3	514	480	346
4	467	840	369
5	513	1300	354

[a] Dioxane was used as solvent for all the measurements;
[b] EFISH experiments were done at a fundamental wavelength 1.907 μm and the $\mu\beta$ values are as measured and not corrected for resonance effect;
[c] Decomposition temperatures were estimated from DSC measurements (Heating rate: 20°C/min).

with thiophene (i.e., compound 5). These $\beta\mu$ values suggest that the loss of activity caused by the replacement of the cyano group in 3 may, to some extent, be compensated for by increasing the conjugation length and introducing stronger donor substituents. These new compounds, despite their higher molecular weight, possess solubility much greater than 2 and this could be used to increase the chromophore number density in polyimide matrices.

DSC studies indicated that the inherent thermal stabilities of compounds 3–5 were on the order of 340–370°C. Unlike the tricyanovinyl derivatives 1 and 2, these modified compounds were totally unreactive to the processing and curing conditions of polyamic acids. Another unusual feature demonstrated by these compounds was their stability to strong bases. For example, treatment with alcoholic KOH resulted in no chemical change in these compounds, while the majority of organic compounds including tricyanovinyl derivatives readily decomposed under these conditions.

In this new series, compound 5 was shown to possess the best optimized optical, thermal and chemical properties and, therefore, was chosen for the electro-optic studies on the polyimide thin-films. Using Ultradel 4212 polyamic acid (solvent: diglyme), polymeric thin-films were formed by spin-coating the combined dye/ polymer solution on an indium tin oxide (ITO) coated glass substrate. The films were kept in a vacuum oven at 120°C for 10 h to remove the solvent. At this stage, the films were imidized by heating at 200°C for 30 min. A thin layer of gold was vacuum evaporated onto the imidized polymer films to serve as the top electrode for poling. The polyimide samples were poled at 220°C for 10 min with an applied dc electric field (1.0 MV/cm). The electro-optic activity of the poled polymer films was measured at 830 nm by an experimental setup similar to that described by Teng and Mann [12]. The measured electro-optic coefficients, r_{33}, were in the range of 10–12 pm/V for polyimide films containing 20 wt % of 5.

To understand the applicability of the material system developed with compound 5 and polyimide PI-4212 in electro-optic devices, further studies such as the index of refraction and attenuation studies were carried out by spin coating the active material on silicon wafers. The index of refraction was measured by the prism coupling method on single layer slabs of core (5/PI-4212) and cladding (PI-4212) materials coated onto 3 μm thermal oxide on silicon wafers. The refractive index difference between the core and cladding as a function of cure temperature is presented in Figure 1. The index difference between the core and clad was observed to be 0.025 and did not change with temperature until 300°C which clearly indicates the high thermal stability of 5 at 300°C. Above 300°C, a drop of index difference was noticed which reflects the out diffusion of the chromophore from the single layer of the guest–host material. This was demonstrated by depositing 0.5 μm of aluminum and performing the same heating sequence, then removing the metal. In this case there was no change in index upon heating to 325°C. Another attractive feature of this material system appeared to be the compatibility between the chromophore and polyimide. This was examined by determining the attenuation of the material system at various cure temperatures in the slab form by the prism coupling and streak method (Figure 2). The attenuation for the fully cured sample was measured, at 830 nm, to be < 2 dB/cm which clearly reflects the compatibility of 5 with the polyimide [13].

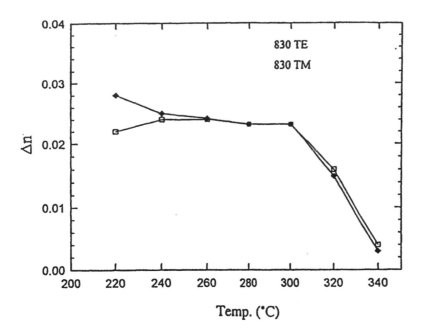

Figure 1 Index of refraction at 830 nm for a guest (5)/host (polyimide 4212) material system *vs.* cure temperature.

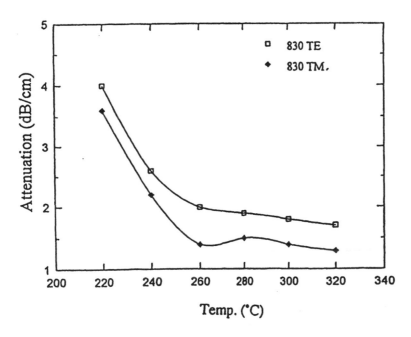

Figure 2 Slab attenuation of guest (5)/host (polyimide 4212) material system *vs.* cure temperature.

The new molecular systems and their properties presented in this text provide insights into the many practical issues relevant to the material development. First of all, the presence of electron-acceptor tricyanovinyl groups and the isomerizable olefinic conjugating bridges cause thermal and chemical instabilities to the chromophoric system. Blocking the most reactive CN group in tricyanovinyl acceptor with aryl group provides unusual chemical and thermal stability at the expense of molecular nonlinearity. Excellent molecular optical, thermal and chemical properties can be achieved with these new chromophores, especially 5.

THERMALLY STABLE POLED POLYQUINOLINE WITH VERY LARGE ELECTRO-OPTIC RESPONSE

The earlier results of design and synthesis of NLO chromophores from Jen et al. [6] and Dirk et al. [14] revealed that using the five-membered heteroaromatic rings as conjugated structures strongly enhance the first order nonlinear optical hyperpolarizability (β). The heteroaromatic chromophore dietheyl-amino-tricyanovinyl substituted cinnamyl thiophene (RT-9800) [6] was employed as dopant in polyquinoline. As reported earlier, this compound demonstrates exceptionally large second order nonlinearity. The $\mu\beta$ value (9800×10^{-48} esu), the product of molecular dipole moment and microscopic second order susceptibility, of this compound measured at 1.907 μm using electric field induced second harmonic generation (EFISH) is approximately 15 times larger than that of diethylamino-nitrostilbene (DANS), which is a commonly used chromophore in second order nonlinear optical material development.

Polyquinolines (Figure 3) are ideal polymer materials for device fabrication because they are mechanically strong, easily processed into low optical loss thin films, and thermally stable up to 500–600°C. Polyquinolines are very soluble in common organic solvents, such as NMP, cyclopentanone, DMAc, etc. The polymer solution can be spin-coated easily onto substrates such as glass and silicon wafers to form thin films. The chromophore RT-9800 is readily dissolved in the polyquinoline/cyclopentanone solution at a 20% by weight content of chromophore relative to the polyquinoline. Further increase of the weight fraction of the chromophore may result in aggregation of the dye. Very good quality thin films (2–4 μm) were obtained by

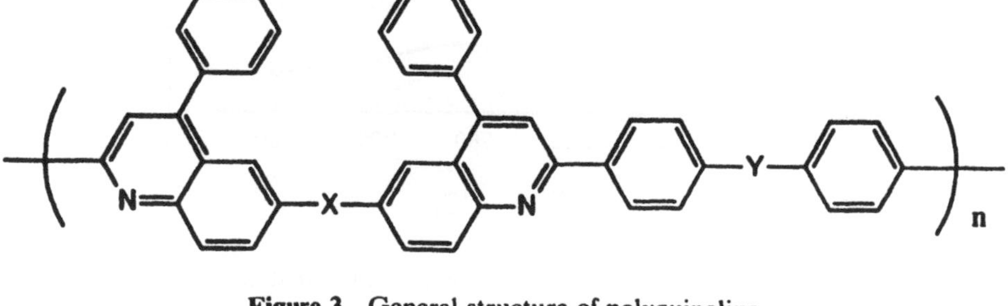

Figure 3 General structure of polyquinoline.

careful spin-coating the filtered (through a 0.2 μm microfilter) polymer solution onto an indium tin oxide (ITO) glass substrate. After being soft baked at 110°C for a few minutes, the films were kept in a vacuum oven for several days at 120°C to remove the residual solvent. A thin layer of gold was then deposited onto the film by sputtering (Desk II, Denton Vacuum). The original glass transition temperature (T_g) of PQ is 265°C, however, after doped with 20 wt% of RT-9800, the T_g dropped to ~ 180°C due to plasticization. The film was poled at 180°C for 5 min in Argon with an applied electric field of around 0.8 MV/cm. The poling field was removed after the chromophore/polyquinoline film was slowly cooled to room temperature.

The electro-optic response of the poled RT-9800/polyquinoline film was measured using an ellipsometric technique described earlier. In order to minimize the contribution due to resonant enhancement, the electro-optic coefficient (r_{33}) was measured at a wavelength of 1.3 μm. For a 20 wt % RT-9800/polyquinoline film poled at 0.8 MV/cm, an r_{33} value of 45 pm/V was achieved [15]. This nonresonant r_{33} value was not only much larger than most of the organic and polymeric nonlinear optical materials, but also exceeds that of LiNbO$_3$, the standard inorganic electro-optic material. Considering the relatively low poling field (less than 1 MV/cm), a higher electro-optic coefficient can be expected if the processing environment can be further improved.

In addition to good electro-optical properties, the ultimate implementation of poled polymer materials in photonic devices requires that the electro-optic activities of the poled films be stable over a long time period at elevated temperatures. In order to test the thermal stability of the poled film, the electro-optic coefficient r_{33} was measured after the sample film was heated isothermally at 80°C over a period of 2000 h. The results are shown in Figure 4. Except for an initial decay of about 40% in the first 100 h, there is little relaxation of the electro-optic activity at a temperature of 80°C. The orientational relaxation phenomena of the electro-optic activity can be empirically described by Kohlrausch–Williams–Watts (KWW) stretched exponential function [16]

$$r_{33}(t) = r_{33}(0)\exp[-(t/\tau)^\beta],$$

where τ is the characteristic time of relaxation and β is a constant between 0 and 1, describing the deviation from the ideal exponential decay process. In fitting our data with this approach, we plotted $\ln[-\ln(r_{33}(t)/r_{33}(0))]$ against $\ln(t)$. From the intercept and slope of the fitting line, we obtain a value of $\beta = 0.085$ and the characteristic time constant $\tau = 1,000,000$ h.

Another alternative for fitting the experimental data is to employ a sum of two exponential functions

$$r_{33}(t)/r_{33}(0) = r_1\exp(-t/\tau_1) + r_2\exp(-t/\tau_2),$$

where $r_1 + r_2 = 1$. Here, τ_1, the "short" characteristic relaxation time is related to the orientational relaxation, and τ_2, the "long" characteristic relaxation time, which is more temperature sensitive, is related to the diffusion of the dipole moments in the

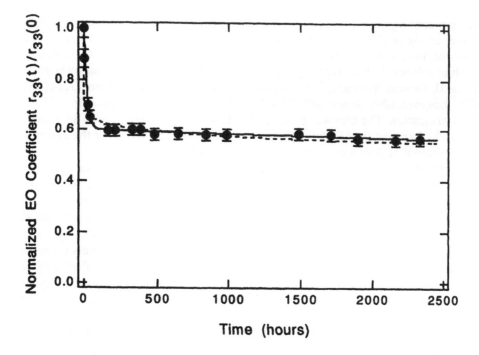

Figure 4 Normalized electro-optic coefficients of RT-9800/polyquinoline at 80°C. The data points are the experiment values. The dotted line is "KWW" stretched exponential function fitting and the solid line is a bi-exponential function fitting.

polymer matrix [17]. We fit our data (Figure 4) using both the "KWW" stretched exponential function (dotted line) and the bi-exponential functions (solid line) using a least-square routine. Since the bi-exponential form produces a better fit result, the "fast" and "slow" characteristic relaxation time approach is a good approximation to describe the complicated, multiplicity of relaxation behavior of the electro-optic activities in this poled polymer thin film. The "short" relaxation time τ_1 and "long" relaxation time τ_2 extracted from the bi-exponential fitting are 20 h and 40,000 h, respectively. The corresponding values of r_1 and r_2 from the curve fitting are 0.41 and 0.59, respectively.

In summary, very large electro-optic activity (45 pm/V) and long-term thermal stability at 80°C for more than 2000 h can be achieved by incorporating hetero-aromatic chromophores with exceptionally large second-order nonlinearities into rigid-rod, high temperature polyquinoline. The experimentally measured r_{33} value agrees relatively well with the theoretical prediction based on EFISH measurements and a two-level model. The long term thermal stability of the electro-optic activity was demonstrated and maintained 26 pm/V at 80°C over a period of more than 2000 h. A bi-exponential function fits our electro-optic relaxation data slightly better than the (KWW) stretched exponential function.

ELECTRO-OPTIC POLYMERS: SIDE-CHAIN POLYIMIDES

Polyimide-based organic materials for second-order nonlinear optics have attracted attention due to their low dielectric constant, high glass-transition temperature (T_g) and compatibility with semiconductor processes [18–21]. The guest–host NLO chromophore-polyimide [18] and aliphatic NLO side-chain polyimide [22–23] are two material systems of polyimides which were developed previously. Although both systems provide encouraging results, their physical properties need to be improved strongly in order to overcome several deficiencies of the systems. Guest–host polyimide system, for example, has several problems, including low loading levels of effective chromophore, sublimation and diffusion of chromophores at high processing temperatures and prolonged working period, plasticization of the polyimide by chromophores, and extraction of chromophore molecules by organic solvents encountered in building multilayer devices [23]; while the aliphatic side-chain polyimide system suffers low glass transition temperature (T_g), low temperature resistance, poor mechanical properties, and the limited structural versatility [22–24]. To overcome these problems, Miller et al. [25], Yu et al. [26] and Jen et al. [27] have recently developed a new NLO polyimide system, NLO side-chain aromatic polyimides. The side-chain aromatic polyimides exhibit much better properties, including higher temperature alignment stability, better mechanical properties, and lower optical loss, than those of aliphatic side-chain and guest–host polyimide systems. However, all these synthetic methods for aromatic side-chain polyimides include a tedious procedure for the synthesis of the chromophore-containing diamine monomers. Moreover, the fact that few chromopores can survive the relatively harsh chemical conditions of the monomer synthesis and the imidization of the polymer severely limits the application of the methodologies [25–27]. Jen et al. had recently developed a facile, generally applicable, two step approach for the synthesis of NLO side-chain aromatic polyimides. This is a one-pot preparation of a preimidized, hydroxy-containing polyimide [28], followed by the covalent bonding of a chromophore onto the backbone of the polyimide via a post-Mitsunobu reaction [29]. By introduction of the chromophores at the last stage through the very mild Mitsunobu condensation, the harsh imidization process of the polyamic acid is avoided and the tedious synthesis of the chromophore-containing diamine monomers is also not necessary. This allows us to synthesize NLO side-chain aromatic polyimides with a broad variation of polymer backbone and almost unlimited flexibility in the selection of the chromophores [30].

The typical procedures for the synthesis of the NLO aromatic polyimides are shown in Scheme 2. Step 1 is for the synthesis of hydroxy-containing aromatic polyimides. The hydroxy-containing diamine, 3,3'-dihydroxy-4,4'-diamino-biphenyl **8** was reacted with a stoichiometric amount of the dianhydride, 2,2'-bis-(3,4-dicarboxyphenyl) hexa-fluoropropane dianhydride **9**. The polymerization was conducted in a closed vessel at 0°C to room temperature over night under nitrogen with a concentration of 15% solids by weight in N,N-dimethyl acetamide (DMAc). The viscosity of the solution increased dramatically during this period. Dry xylene was added into the flask and the polyamic acid was thermally cyclized at 160°C for 3 h.

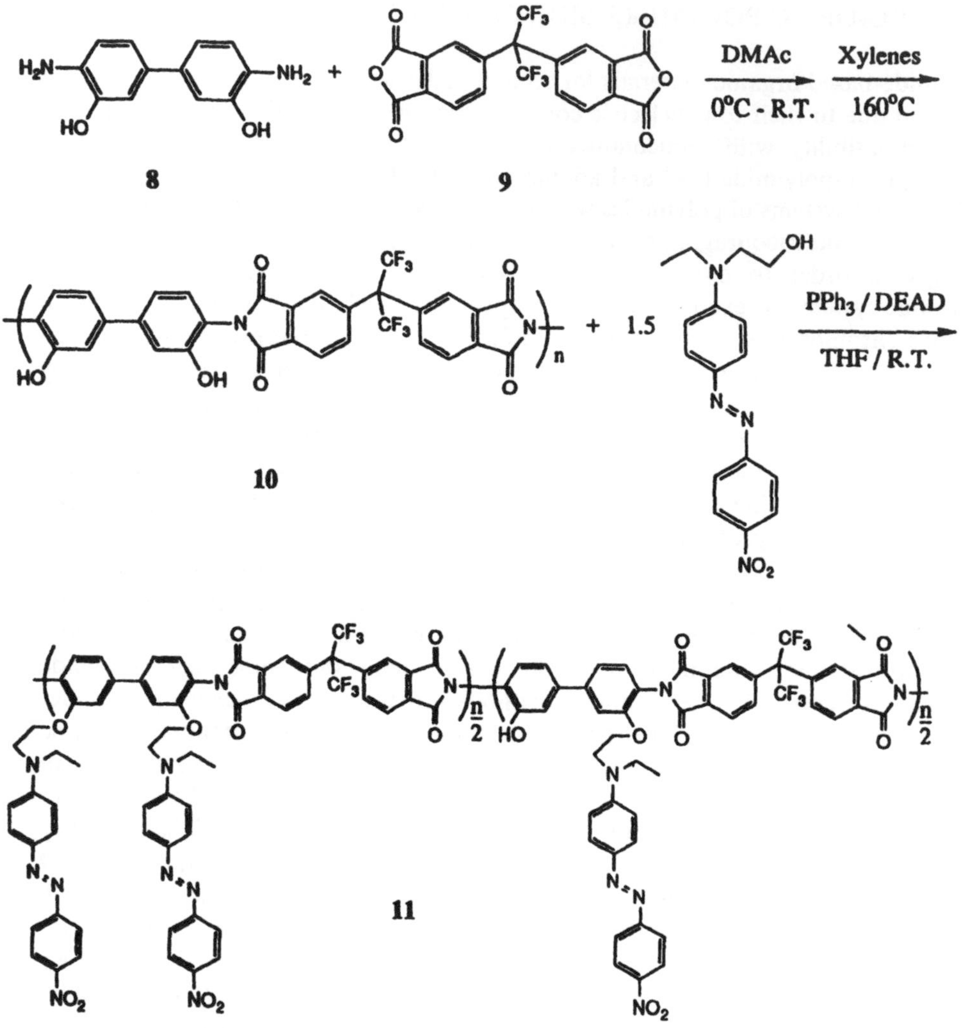

DEAD: Diethyl azodicarboxylate; DMAc: N, N - Dimethyl acetamide

Scheme 2

Water that was eliminated by the ring closure reaction was removed as a xylenes azeotrope at the same time. The resulting solution was added dropwise into an agitated solution of methanol (500 mL) and 2 N HCl (10 mL) to obtain the brown color hydroxy-polyimide 10. The polymer 10 was further purified by reprecipitation into a solution of methanol (500 mL) and 2 N HCl (10 mL) from its THF (50 mL) solution. The solid was dried at 60°C under vacuum for 24 h to produce polyimide 10 with a yield of 91%. Step 2 of the synthesis is for the covalently bonding of the hydroxy-containing chromophore onto the back of the polyimide **10**. Disperse red 1 (0.8 equivalents, relative to the equivalent of the hydroxy groups of polymer **10**) was reacted with polymer **10** via Mitsunobu condensation to produce NLO side-chain aromatic polyimide **11** in 92% yield.

The polymerization of polyimide **10** and the Mitsunobu condensation between **10** and disperse red were monitored by proton NMR spectroscopy. The ^{1}H NMR (DMSO-d$_6$) spectrum of **10** shows a completely imidized polyimide structure with the aromatic hydroxy resonance at 10.06 ppm. The three sets of resonance peaks at 8.20, 8.00, and 7.80 ppm were due to the aromatic protons of the 6F-dianhydride moieties. Another three sets of peaks at 7.39, 7.21, and 7.18 ppm were attributed to the aromatic protons of the dihydroxy-biphenyl moieties. The ^{1}H NMR (DMSO-d$_6$) of polyimide **11** shows that approximately 75% of the – OH in polymer **10** reacted with disperse red and the other 25% was intact. The chemical shift of the intact – OH protons was observed at 10.02 ppm. The resonances between 6.6 and 8.4 ppm belong to all the aromatic protons of the backbone and the side-chain chromophore of polymer **11**. The resonances of the side-chain aliphatic protons of **11** appeared at 4.37, 3.75, 3.26, and 0.88 ppm.

The post-Mitsunobu condensation between hydroxy polyimides, such as **10**, and hydroxy chromophores, such as disperse red, was surprisingly quantitative, which allowed us to adjust the loading level of the side-chain chromophores and the density of the intact OH groups efficiently. The hydroxy-containing side-chain NLO polyimides can be further crosslinked to different extents in order to improve the mechanical roperties, solvent resistance, and thermal alignment stability of the materials [31].

The NLO polyimide was soluble in polar solvents such as cyclohexanone, DMAc, NMP, as well as THF. The molecular weights of the polymer can thus be estimated by gel permeation chromatography (GPC). Polymer **11** has a weight-averaged molecular weight (relative to polystyrene standard) M_w of 56000 with a polydispersity index of 1.44. The polymer also has a high T_g and excellent thermal stability. Although the chromophore loading level is as high as 42% by weight, the polymer **11** still has a T_g of 228°C (by DSC) and a thermal stability of < 1% weight loss up to 300°C (by TGA analysis). The UV-vis spectrum of thin film of **11** exhibited a strong absorption pattern ($\lambda_{max} = 480$ nm) due to the π–π^* charge-transfer band of the side-chain disperse red.

Optical quality thin film ($\sim 3\,\mu$m) of polymer **11** was prepared by spin-coating of the polymer solution in cyclohexanone (15% m/m, filtered through 0.2 μm syringe filter) onto an indium tin oxide (ITO) glass substrate. The dipole alignments in the NLO polymas can be achieved and the second-order nonlinearity can be induced either by contact poling or corona poling. Preliminary tests show that the synthesized NLO side-chain polyimides exhibit a large E-O coefficient (r_{33}) and good thermal stability of dipole alignment. The preliminary r_{33} value for polymer **11** was 30 pm/V measured at 0.63 μm and 10 pm/V measured at 0.83 μm. The r_{33} value retained > 95% of the original value at 85°C and > 90% at 100°C for more than 400 h. The further optimization of the E-O coefficient and the thermal alignment stability of the NLO polyimides and their crosslinking system are currently under investigation.

In summary, a two-step and generally applicable synthesis for NLO side-chain aromatic polyimides was developed. A hydroxy-containing NLO polyimide with the disperse red 1 as the side-chain group was synthesized and characterized. The resulting NLO polyimide possesses excellent solubility and processibility even though the loading level of the side-chain chromophore is up to 42% by weight. A large E-O coefficient value of 30 pm/V (at 0.63 μm) and 10 pm/V (at 0.83 μm) was achieved and a high thermal stability of the poled film at 100°C was observed. We are continuing to examine many aspects which may further improve the properties of the NLO aromatic polymeric materials.

REFERENCES

[1]   Hornak, L. A. (ed.), *Polymers for Lightwave and Integrated Optics*, Marcel Dekker, New York (1992).

[2]   Chemala, D. S. and Zyss, J. (eds.), *Nonlinear Optical Properties of Organic Molecules and Crystals*, Academic Press, New York (1987).

[3]   Lytel, R. and Lipscomb, G. S., *Mater. Res. Soc. Symp. Proc.* **247** (1992) 17.

[4]   (a) Marder, S. R., Sohn, J. E. and Stucky, G. D. (eds.), Materials for Nonlinear Optics Perspectives, *ACS Symposium Series 455*, Washington (1991); (b) Garito, A. F., Jen, A. K.-Y., Lee, C. Y.-C. and Dalton, L. R. (eds.), Electrical, Optical and Magnetic Properties of Organic Solid State Materials, *Mater. Res. Soc. Symp. Proc.* **328** (1994).

[5]   Drost, K. J., Jen, A. K.-Y. and Rao, V. P., *ChemTech* **25** (1995) 16.

[6]   Jen, A. K.-Y., Wong, K. Y., Rao, V. P., Drost, K. J. and Cai, Y., *J. Electronic Mater.* **23** (1994) 653.

[7]   Rao, V. P., Jen, A. K.-Y., Wong, K. Y. and Drost, K. J., *Tetrahedron Lett.* **34** (1993) 1747.

[8]   Jen, A. K.-Y., Rao, V. P., Wong, K. Y. and Drost, K. J., *J. Chem. Soc. Chem. Commun.* **90** (1993).

[9]   Rao, V. P., Jen, A. K.-Y., Wong, K. Y. and Drost, K. J., *J. Chem. Soc. Chem. Commun.* (1993) 1118.

[10] Rao, V. P., Cai, Y. and Jen, A. K.-Y., *J. Chem. Soc. Chem. Commun.* (1994) 1689.

[11] Wong, K. Y. and Jen, A. K.-Y., *J. Appl. Phys.* **75** (1994) 3308.

[12] Teng, C. C. and Man, H. A., *Appl. Phys. Lett.* **56** (1990) 1754.

[13] Kenney, J. T., Binkley, E. S., Jen, A. K.-Y. and Wong, K. Y., *Mater. Res. Soc. Symp. Proc.* **328** (1993) 511.

[14] Dirk, G. W., Katz, H. E., Schilling, M. L., Singer, K. D. and Sohn, J. E., *Chem. Mater.* **2** (1990) 700.

[15] Cai, Y. M. and Jen, A. K.-Y., *Appl. Phys. Lett.* **67**(3) (1995) 299.

[16] Singer, K. D. and King, L., *J. Appl. Phys.* **70** (1991) 3251.

[17] Hampsch, H., Yang, J., Wong, G. and Torklson, J., *Macromolecules* **23** (1990) 3640.

[18] Wu, J., Valley, J. F., Ermer, S., Binkley, E. S., Kenney, J. T., Lipscomb, G. F. and Lytel, R., *Appl. Phys. Lett.* **58** (1991) 225.

[19] Xu, C., Wu, B., Dalton, L. R., Shi, Y., Ranon, P. M. and Steier, W. H., *Macromolecules* **24** (1991) 5421.

[20] Park, J., Marks, T., Yang, J. and Wong, G. K., *Chem. Mater.* **2** (1990) 229.

[21] Zysset, B., Ahlheim, M., Stahelin, M., Lehr, F., Pretre, P., Kaatz, P. and Gunter, P., *Proc. SPIE* **2025** (1993) 70.

[22] Becker, M., Sapochak, L., Ghosen, R., Xu, C., Dalton, L. R., Shi, Y., Steier, W. H. and Jen, A. K.-Y., *Chem. Mater.* **6** (1994) 104.

[23] Peng, Z. and Yu, L., *Macromolecules* **27** (1994) 2638.

[24] Jen, A. K.-Y., Drost, K. J., Cai, Y., Rao, V. P. and Dalton, L. R., *J. Chem. Soc. Chem. Commun.* (1994) 965.

[25] (a) Miller, R. D., Burland, D. M., Dawson, D., Hedrick, J., Lee, V. Y., Moylan, C. R., Twieg, R. J., Volksen, W. and Walsh, C. A., *Polym. Prepr.* **35** (1994) 122.

[26] (a) Yu, D., Gharavi, A. Yu, L *Macromolecules* **28** (1995) 784; (b) Yu, D. and Yu, L., *Macromolecules* **27** (1994) 6718.

[27] (a) Jen, A. K.-Y., Liu, Y. J., Cai, Y., Rao, V. P. and Dalton, L. R., *J. Chem. Soc. Chem. Commun.* (1994) 2711.

[28] Jen, A. K.-Y., Cai, Y., Drost, K. J., Liu, Y. J., Rao, V. P., Chen, T.-A. and Kenney, J. T., *Proceedings of the Am. Chem. Soc., PMSE* **72** (1995) 213.

[29] Mitsunobu, O., Synthesis, 1981, 1.

[30] Chen, T.-A., Jen, A. K.-Y. and Cai, Y., *J. Am. Chem. Soc.* **117** (1995) 7295.

[31] Marks, T. J. and Ratner, M. A., *Angew. Chem. Int. Ed. Engl.* **34** (1995) 155.

15. FREQUENCY DOUBLING WITH NONLINEAR OPTICAL POLYMERS

K. SASAKI, S. KIM, G. J. ZHANG, and S. HORINOUCHI

*Department of Material Science, Keio University,
3-14-1 Hiyoshi, Yokohama 223, Japan*

ABSTRACT

Disperse Red 1 (DR1) side chain copolymer with polymethyl methacrylate (PMMA) was prepared for poled polymer films which were categorized into seven frequency doubling waveguide devices on linear and nonlinear substrates with Cerenkov-type phase matching. In the categorization the most sophisticated structure was the face-to-face combined structure of two waveguides on nonlinear potassium di-hydrogen phosphate (KDP) crystal substrates with counter polarity in quadratic coefficient which gave the highest theoretical conversion efficiency in the categorization.

1. INTRODUCTION

One of the most potential application of organic nonlinear molecular is guest–host system such as side-chain pendants of nonlinear molecules adding to main chain of polymer system. The system usually is thermally stable and good in processability for preparation of film. In this study disperse red 1 (DR1) was used as pendant molecule to polymethylmethacrylate (PMMA) main chain polymer. The material system was applied to nonlinear waveguide films on linear and nonlinear substrates with poling processing. Waveguides on linear and nonlinear substrates were categorized into seven types of waveguides by cross-sectional structures with quadratic polarity of nonlinear coefficient. Cross-sectional field distributions of the fundamental and the second harmonic waves give the conversion efficiency of the second harmonic generation via the evaluation of overlap integral.

2. PREPARATION OF BASIC WAVEGUIDE

Azo dye, disperse red 1 (DR1) substituted methyl methacrylate (MMA) and MMA monomer were synthesized as the host material as shown in Figure 1.

A typical host monomer of molar fraction, $x = 0.023$ (sample, #2) was spin-coated on a pyrex substrate with appropriate viscosity. Corona poling was applied to the alignment of DR1 side chain molecules with de-electric field (5 kV/cm) at 130°C. Measured d_{33} value by Maker fringe method was 9.0 pm/v. Also measured dispersion curves are shown in Figure 2 for two typical samples (#2, #5) together with Table 1.

Two basic waveguide structures were prepared: (1) poled polymer waveguide on optically linear pyrex substrate, (2) poled polymer waveguide on optically nonlinear potassium di-hydrogen phosphate (KDP) single crystal substrate.

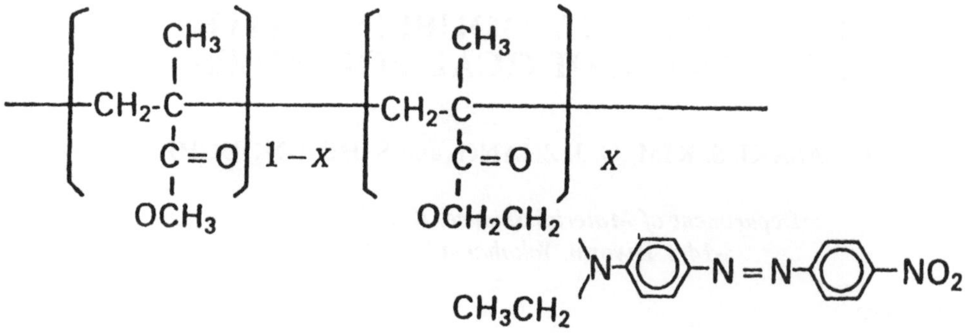

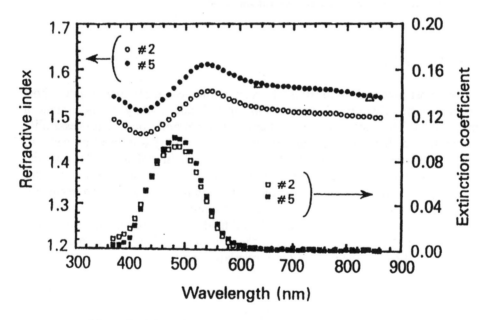

Figure 1 Copolymer of Poly (MMA-co-DR1MA).

Figure 2 Dispersion curves of DR1/PMMA copolymer.

Table 1 Measured d_{33} values and refractive indices.

Sample#	Molar fraction	d_{33} (pm/v)	n (633 nm)	n (840 nm)	n (1064 nm)
#2	0.023	9.0	1.509	1.501	1.459
#5	0.073	37.5	1.567	1.539	1.529

3. CALCULATIONS ON CONVERSION EFFICIENCIES OF FREQUENCY DOUBLING FOR CATEGORIZED WAVEGUIDE STRUCTURES

Two basic waveguides were categorized into seven kinds of cross-sectional structures. Waveguides on nonlinear KDP substrates were specified by the sign of polarities between the waveguide and nonlinear KDP substrate. That is characterized by the arrow as shown in Figure 3:

(a) nonlinear waveguide on linear pyrex substrate with open face asymmetrical structure,

(b) open face asymmetrical nonlinear waveguide on nonlinear KDP substrate with parallel polarity,

(c) open face asymmetrical nonlinear waveguide on nonlinear KDP substrate with counter polarity,

(d) face-to-face symmetrical waveguide on linear pyrex substrate with parallel polarity,

(e) face-to-face symmetrical waveguide on linear pyrex substrate with counter polarity,

(f) face-to-face symmetrical counter polarity waveguide on nonlinear counter parallel KDP substrate,

(g) face-to-face symmetrical parallel polarity waveguide on nonlinear counter parallel KDP substrate.

As is well known the overlap integral in the conversion efficiency was given via cross-sectional field distribution of waveguide. Used parameters in calculations of overlap integrations for seven waveguide structures are given in Table 2.

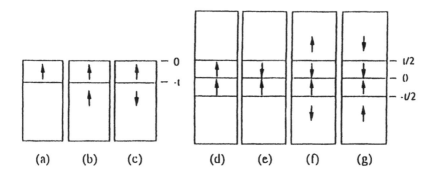

Figure 3 Categorized waveguide structures: (a) nonlinear waveguide on linear pyrex substrate with open-face asymmetrical structure, (b) open-face asymmetrical nonlinear waveguide on nonlinear KDP substrate with parallel polarity, (c) open-face asymmetrical nonlinear waveguide on nonlinear KDP substrate with counter polarity, (d) face-to-face symmetrical waveguide on linear pyrex substrate with parallel polarity, (e) face-to-face symmetrical waveguide on linear pyrex substrate with counter polarity, (f) face-to-face symmetrical counter polarity waveguide on nonlinear counter parallel KDP substrate, (g) face-to-face symmetrical parallel polarity waveguide on nonlinear counter parallel polarity.

Table 2 Used parameters.

	n (532 nm)	n (1064 nm)	
Pyrex	1.476	1.464	
DR1/PMMA	1.600	1.534	$d_{33} = 10.0\,\text{pm/v}$
KDP n_o	1.512	1.501	$d_{36} = 0.41\,\text{pm/v}$
n_e	1.471	1.461	

Interaction length: 10 mm
Fundamental input: 1 W
Beam width: 5 micron

Anisotropic refractive indices and quadratic nonlinear coefficients of KDP crystal were considered. Comparison of conversion efficiencies for the seven structure is shown in Figure 4. In the figure, the structure (f) gives the highest conversion efficiency with Cerenkov-type phase matching. From the result the face-to-face counter parallel waveguide on nonlinear substrates is advantageous for efficient frequency doubling.

Effective cross-sectional field distributions and phase matching curves are shown in Figures 5 and 6, respectively.

Two cases, (f) and (g) both are indicated in Figure 5, the upper indication means the counter polarity waveguide on counter polarity nonlinear substrates, (f) and the

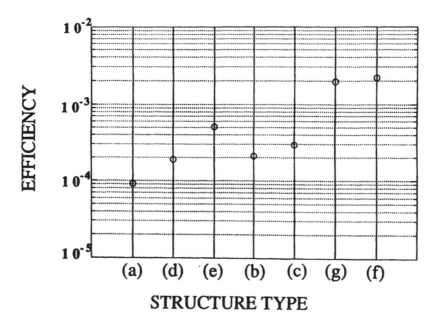

Figure 4 Conversion efficiencies of categorized waveguides.

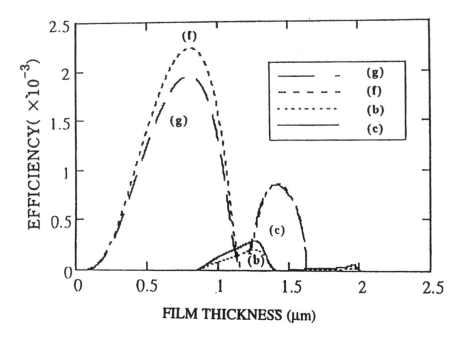

Figure 5   Field distributions of (f) and (g) types waveguides.

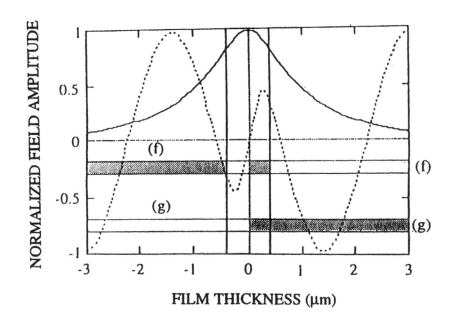

Figure 6   Thickness dependent conversion efficiencies.

lower illustrates the counter polarity waveguides on the parallel polarity nonlinear substrates. In Figure 6 the structure (f) gives the highest conversion efficiency with rather wide phase matching tolerance in waveguide thickness.

CONCLUDING REMARKS

DR1/PMMA copolymer film waveguides on linear pyrex substrate and on nonlinear KDP crystal substrate are categorized by combinations on nonlinear polarities and cross-sectional structures. Waveguides gave frequency doubling in Cerenkov-type phase matching. The categorization (f) gave the highest conversion efficiency in seven types of waveguide in Figure 3. This kind of concept can be applicable to the high efficiency collinear phase matched frequency doubling waveguide.

ACKNOWLEDGEMENT

The authors would like to express their thanks to Professor Takatomo Sasaki, Osaka University for his support on nonlinear crystals.

16. SHG OF ORGANIC PHOTOCHROMES IN POLYMER MATRIX: PHOTO-ASSISTED POLING AND PHOTO-SWITCHING

KEITARO NAKATANI, YOMEN ATASSI,
and JACQUES A. DELAIRE

Laboratoire de Photophysique et de Photochimie Supramoléculaires et Macromoléculaires (C.N.R.S. U.R.A. 1906), Ecole Normale Supérieure de Cachan, 61 avenue du Président Wilson, 94235 Cachan Cedex, France

ABSTRACT

Polymers containing molecules exhibiting both quadratic nonlinear optical properties and photochromism were studied. The combination of these two properties within the same polymer material leads to two phenomena. On the one hand, when the photochemical transformation is performed on an isotropic sample under a static electric field at room temperature, a noncentrosymmetric order can be achieved. Susceptibility values (d_{33}) of 13 pm/V and 35 pm/V could be reached respectively with Disperse Red One and spiropyran-photomerocyanine. This phenomenon has also been observed with the furylfulgide-dihydrobenzofuran photochromic system. Moreover, a synergistic effect of the electric field and the photochemical process has been evidenced for the spiropyran-photomerocyanine system. On the other hand, nonlinear optical properties can be switched by this light induced molecular change, though total reversibility has not been reached yet. In this paper, we also discuss about the potential applications of materials exhibiting these phenomena.

INTRODUCTION

Polymers as optoelectronic materials are getting widely studied, both from fundamental and practical aspects [1–3]. Contrarily to crystalline systems, these materials can be easily processed as thin films, and introduced as device components. Their relatively low cost is also another advantage. Moreover, the flexibility and the almost limitless possibilities offered by molecular chemistry enable us to build up doped or functionalized molecular-based materials with various properties, making them more competitive. In this respect, our study focuses on polymer matrices combining two different properties, namely nonlinear optical (NLO) properties and photochromism. Photochromic compounds have been studied for many years and are still being thoroughly investigated for their potential applications in glasses, lenses, memories and switching materials [4–8]. Among all external perturbations that yield a molecule's and material's property change, those induced by light seem to be worth studying for practical reasons. Light induced NLO property change has been suggested and studied on a molecular scale by Lehn and co-workers [9]. Furthermore, as it is well known, the orientation of molecules within the polymer matrix has a large influence on the NLO properties, and the noncentrosymmetry of the material is even a prerequisite to have quadratic NLO effects. Molecular structure change cannot be dissociated from orientational change on a macroscopic scale, and this has been studied by several research groups throughout the world [10,11]. Our study highlights the combination of photochromism and quadratic NLO properties, which leads to two phenomena: on the one hand, photoassisted poling at room temperature in presence of an external electric field; and on the other hand, the photoswitching of NLO properties of a polymer sample.

PHOTOCHROMIC MOLECULES AND THEIR NLO PROPERTIES

Before developing the different phenomena, let us describe the photochromic systems and their physical properties.

Photochromic molecules exhibit reversible molecular transformation when excited with radiation. Usually both forms **a** and **b** show distinct absorption spectra. Reaction from **b** form to **a** form may also occur thermally in the absence of any radiation in some photochromic systems. Our study focused on three families of photochromes (Table 1) [12,13]:

$$a \underset{h\nu'}{\overset{h\nu}{\rightleftharpoons}} b$$

Table 1   Photochromic molecules.

a	Thermal back reaction	b
trans Disperse Red One (1a) N-ethyl-N-(2'-hydroxyethyl) ((4-nitrophenyl)azo) aniline) or DR1	Yes	cis Disperse Red One (1b)
Spiropyran (2a) 6-nitro-1',3',3'-trimethylspiro [2H-1-benzopyran-2,2'-indoline] or 6-nitro-BIPS	Yes	Photomerocyanine (2b)
Furylfulgide (3a) (E)-α-(2,5-dimethyl-3-furyl-ethylidene)(isopropylidene) succinic anhydride or Aberchrome 540	No	Dihydrobenzofuran (3b)

When irradiated at an appropriate wavelength, all these molecules undergo photo-chromic reactions in polymethylmethacrylate (PMMA) matrix. These were followed by measuring the optical density (OD) change of PMMA thin films. For systems 2 and 3, UV–visible absorption spectra before and after irradiation in the UV region (355 nm) were compared (Figures 1 and 2). Though unstable at room temperature, we could determine the absorption spectrum of 1b (Figure 3), combining methods described by Rau [14] and Fischer [15]. The quantum yields have been determined for photochromic reactions in 1 and 2 (Table 2). These are rather significant and efficient conversion between forms a and b is achieved. The three photochromes are different in the way that for 1 and 2, molecular form b (the less stable one) can

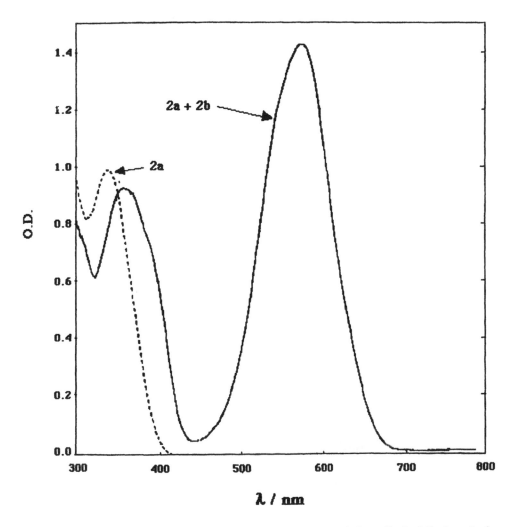

Figure 1 UV–visible absorption spectra of 2a (spiropyran) before (dashed line) and after irradiation at 355 nm (full line) in PMMA film (10% w/w).

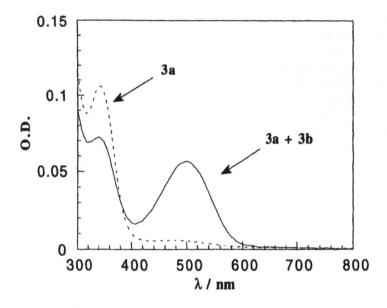

Figure 2 UV–visible absorption spectra of **3a** (furylfulgide) before (dashed line) and after irradiation at 355 nm (full line) in PMMA film (10% w/w).

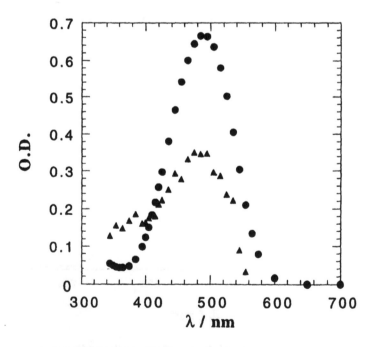

Figure 3 UV–visible absorption spectra of **1a** (trans-DR1, ●) and **1b** (cis-DR1, ▲) in PMMA film (5% w/w) determined by Fischer's method.

Table 2 Quantum yields ($\Phi_{a \to b}$ and $\Phi_{b \to a}$) and rate constants (k_1 and k_2) for the thermal back reaction for **1** (DR1) and **2** (spiropyran-photo-merocyanine); the kinetics was approximated by a monoexponential for **1** and by a biexponential for **2**.

Compound	$\Phi_{a \to b}$	$\Phi_{a \to b}$	k_1/s^{-1}	k_2/s^{-1}
1 DR1 (trans-cis)	0.11	0.7	5.6	
2 Spiropyran-photomerocyanine	0.10	0.02	$7 \cdot 10^{-4}$	$1 \cdot 10^{-4}$

convert back to molecular form **a** both photochemically and thermally, whereas in **3**, this reaction can only be light induced [16]. Though the thermal reactions in **1** and **2** do not follow an exponential law, the comparison of the fastest component of these kinetics show that this reaction is about 10^4 times faster in **1** than in **2**.

In all these systems, there is a rather important change of the ground state dipole moment (μ) and of the quadratic hyperpolarizability (β). β values were determined by finite field method (Table 3). The delocalization of π electrons in **2b** and **3b** explains the high values of μ and β in these molecules. It is worth mentioning that, in **2b**, μ and β have opposite signs, which means that the charge transfer is expected to have opposite directions in the ground state and the excited one. This opposite sign between μ and β has already been observed in merocyanine type molecules [17].

EXPERIMENTAL SETUP

Polymer samples of photochromic polymers were prepared by spin-coating technique. Molecules were dissolved in their **a** form in a suitable solvent (usually toluene or chloroform) along with PMMA [12,13]. **1a** and **2a** are available from Aldrich Co., **3a** from Aberchromics Ltd., and PMMA (Elvacite 2041) from DuPont. Concentrations of chromophore ranged from 5% to 25% w/w relative to polymer. In some of

Table 3 Comparison of dipole moments (μ, upper line) and quadratic hyper-polarizabilities (β, lower line) between molecular forms **a** and **b**; β values correspond to the projection of the vectorial components of the quadratic susceptibilities on μ.

Compound	a form	b form
1 DR1 (trans-cis)	7.7 D	—
	$45 \cdot 10^{-30}$ esu	$8.4 \cdot 10^{-30}$ esu
2 Spiropyran-photomerocyanine	4.5 D	19.2 D
	$1.0 \cdot 10^{-30}$ esu	$-40 \cdot 10^{-30}$ esu
3 Furylfulgide-dihydrobenzofuran	7.2 D	6.6 D
	$0.8 \cdot 10^{-30}$ esu	$6.0 \cdot 10^{-30}$ esu

the experiments, samples were covered by a gold electrode. For photoassisted poling experiments, polymer films were not poled prior to the NLO experiments, whereas for photoswitching experiments, they were poled by the thermally assisted poling, i.e. by heating the sample up to its glass transition temperature (T_g) and cooling it down to room temperature within an electric field.

NLO properties of the materials were studied by second harmonic generation (SHG) technique. These were performed on a Nd/YAG laser (pulsewidth 30 ps, repetition rate 10 Hz). The fundamental beam at 1064 nm was used to probe the sample by SHG, whereas simultaneously another laser beam pumped the photochromic molecules to yield reaction (Figure 4) [13]. Depending on the molecule studied, we used different types of pump beams: the third harmonics of the Nd/YAG laser (355 nm, 1 mJ) for photochromes **2a** and **3a**, and an argon ion CW laser (488 nm, 1 mW) for photochrome **1a**. For photoswitching experiments, a second pump beam was used, in our case the 514 nm beam (1 mW) of the argon ion CW laser. For photoassisted poling experiments, either a voltage of 200 V was applied to the gold electrode on the film surface, or a voltage of 5 kV was applied to a metal needle (corona needle) positioned at a distance of 1 cm from the film surface, in order to create a poling field (E_p) within the polymer sample by corona discharge. For the determination of d_{33}, the SHG signal generated by the polymer was compared to the signal generated by a reference quartz sample. All these NLO experiments were performed at room temperature.

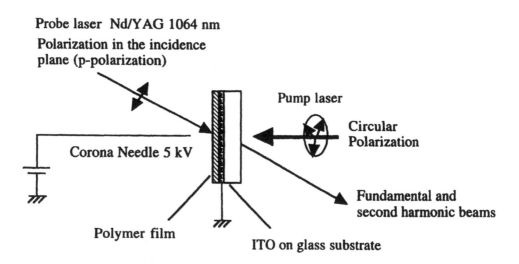

Figure 4   Experimental setup for *in situ* measurements of SHG (configuration around the sample). Corona discharge by the needle can be replaced by a voltage (200 V) applied directly on a gold electrode coating on the film surface. Argon ion CW laser (488 nm or 514 nm) or the third harmonics (355 nm) of the Nd/YAG laser serve as pump beam.

PHOTOASSISTED POLING

Disperse Red One (DR1, 1a–1b)

Photoassisted poling was performed on DR1-doped (host–guest) and -functionalized (side chain) PMMA matrix on the experimental setup described previously. This consisted in applying an external electric field E_p and in irradiating the sample, while monitoring the SHG signal ($I_{2\omega}$). The SHG signal is represented in Figure 5 [18].

First, the isotropic sample of **1a** (trans-DR1) does not show any SHG. The application of E_p by corona discharge evolves an SHG signal. This is probably due to the $\chi^{(3)}$ contribution to SHG when a static electric field is applied. In fact, even if no orientation occurs and $\chi^{(2)} = 0$, there is an SHG signal in presence of E_p:

$$I_{2\omega} \alpha [(\chi^{(2)} + \chi^{(3)} E_p) E_\omega E_\omega]^2.$$

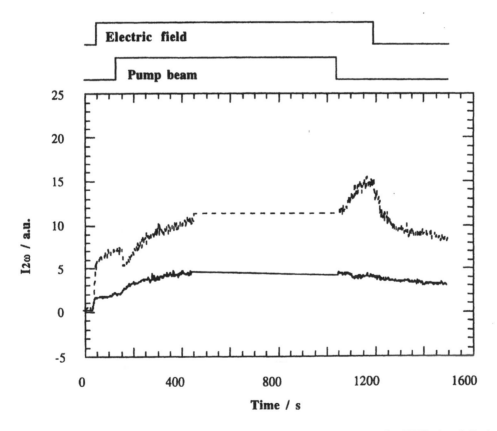

Figure 5 Photoassisted poling at room temperature monitored *in situ* by SHG signal ($I_{2\omega}$) measurement on **1** (DR1, 5% w/w doped polymer, full line and 23% mol/mol functionalized, dashed line).

After the application of the pump beam, $I_{2\omega}$ shows a slow increase, as molecules tend to align with the electric field. In the case of the functionalized system, a fast drop of the signal occurs before this increase, due to a lowering of E_p because charge injection occurs in presence of light. When the pump beam is switched off, a sharp increase occurs in the case of the functionalized system, which corresponds to the recovery of E_p. The highest value of $I_{2\omega}$ reached in this experiment corresponds to a value of $d_{33} = 13\,\text{pm/V}$. This value is close to that obtained by thermally assisted poling. A study by Blanchard and Mitchell compares both poling methods in the case of the doped polymer [19]. After the removal of E_p, $I_{2\omega}$ slowly decreases by charge removal in the first minutes, and by slow disorientation on a longer timescale.

This phenomenon can be schematized (Figure 6) and explained as follows: light induces the trans → cis → trans photoisomerization of DR1 molecules, exciting with higher probability those whose transition moment is in the plane of the film. Consequently, after repeating photoisomerization cycles, DR1 molecules tend to line up along the direction perpendicular to the plane of the film. When light excitation is applied alone, anisotropy is created but the material remains centrosymmetric, whereas when light excitation and electric field are both applied, noncentrosymmetric poling occurs. Two effects are believed to contribute to this process: first, **1b** (cis-DR1) is more globular and more mobile than its isomer; second, during the isomerization, electric field may bias the reorientation of molecules preferentially along the field direction. This phenomenon has been put into evidence by attenuated total reflectance (ATR) experiments by Dumont and co-workers [20,21].

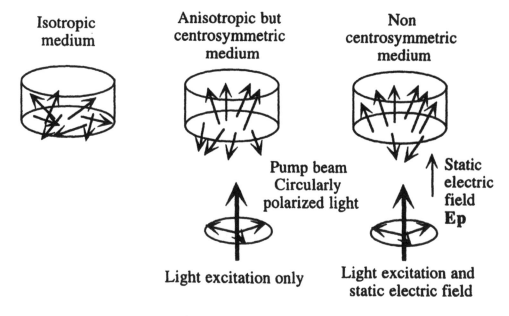

Figure 6 Principle of photoassisted poling: a schematic view; when light is applied alone, the medium becomes anisotropic but remains centrosymmetric whereas when light and electric field are both applied, the medium becomes noncentrosymmetric.

Spiropyran-Photomerocyanine (2a–2b)

Since the thermal back reaction in this system is about 10^4 times slower than in **1**, NLO properties of the less stable molecular form **b** can be investigated. Moreover, contrarily to **1** where **1b** has a lower β value than **1a**, **2b** has a higher β value than **2a**.

PMMA films doped with **2a** were prepared and covered with a gold electrode. Two types of experiments were performed on identical films (Figure 7). Experiment I consists in applying simultaneously both the pump beam and the electric field E_p whereas Experiment II consists in applying E_p once **2a** has been transformed to **2b**. The main feature is the appearance and the increase of $I_{2\omega}$ when E_p is applied in Experiment II, and when the pump beam is switched on in Experiment I. In both experiments, a plateau value of $I_{2\omega}$ is reached after a few minutes. This value is higher

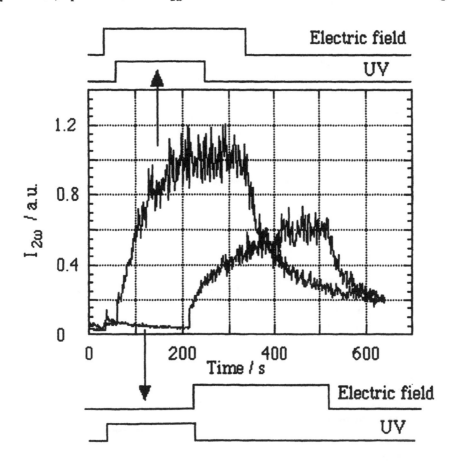

Figure 7 Photoassisted poling at room temperature monitored *in situ* by SHG signal ($I_{2\omega}$) measurement on **2** (spiropyran-photomerocyanine, 25% w/w): Experiment I (simultaneous application of UV irradiation and electric field E_p, upper curve) shows more efficient poling than Experiment II (UV irradiation applied previously to E_p, lower curve) under the same pumping and poling field conditions.

in Experiment I. Though the ratio of the $I_{2\omega}$ values reached between the two experiments fluctuates from 1.5 to 3, a more efficient SHG is obtained in Experiment I than in Experiment II. Since irradiation and electric field application conditions are exactly the same in both experiments, we assume that the difference of SHG intensity is mainly due to a different orientation efficiency. This phenomenon has also been observed for other spiropyran-photomerocyanines [22]. If we make a closer analysis of the kinetics of the signal growth, $I_{2\omega}$ increases like the kinetics of the rotational diffusion of dipoles under an electric field in Experiment I. The increase of $I_{2\omega}$ in Experiment I can be compared with the rate of the ring opening of the molecules. It is also worthwhile to note that if we wait a few minutes between both perturbations in Experiment II, the plateau value reached is lower. Since the amount of **2b** that disappears in the dark is negligible at that time scale, this can be attributed to a reorganization of the polymer matrix around the **2b** (photomerocyanine) molecules, which prevents efficient orientation when the electric field is not applied immediately after ring opening reaction. All these results have also been observed by ATR experiments by Dumont and co-workers [23].

A model based on rotational diffusion of molecules in the polymer matrix is currently being investigated to rationalize the difference between Experiment I and Experiment II. Up to now, a comparison between data from Experiment I and those obtained by a convolution product of data from Experiment II by the kinetics of **2a** → **2b** reaction has been made. If we neglect the contribution of **2a** to SHG, $I_{2\omega}$ is proportional to N^2, N being the number of **2b** molecules. Therefore, the square root of the SHG signal in Experiment I, $I_1^{1/2}$, was compared to $I_1'^{1/2}$ defined as follows (Figure 8):

$$I_1'^{1/2} = \int_0^t \left(\frac{dN}{dt}\right)_{t=\tau} \left[I_2(t-\tau)\right]^{1/2} d\tau.$$

This comparison has been made for two different pump intensities. In both cases, $I_1^{1/2}$ has a faster growth than $I_1'^{1/2}$, which agrees with the fact that the orientation of the molecules occur both during and after isomerization. This highlights the synergistic effect of light induced transformation and electric field.

In both experiments, SHG signal decays after switching off the electric field E_p. On the time scale of the experiment, the ring closure of **2b** to **2a** is not of significant extent. Hence, this signal decrease is mainly due to a disorientation mechanism that fits a Kohlrausch–Williams–Watts stretched exponential law ($I_{2\omega} = I_{2\omega}(0) \exp[(-t/\tau)^\alpha]$ with $\tau = 74\,\text{s}$ and $\alpha = 0.29$) [24,25].

Furylfulgide-Dihydrobenzofuran (3a–3b)

The same experiments have been performed on this photochrome. The only difference with the preceding experiments is the use of a corona discharge instead of a gold electrode. $I_{2\omega}$ exhibits a very sharp increase when the electric field E_p is applied after UV irradiation (Figure 9). Contrarily to **2a** that yields a ring opening

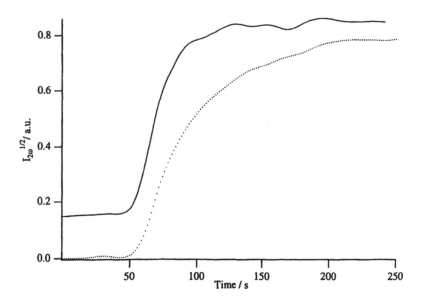

Figure 8 Convolution product from Experiment II with the kinetics of **2a → 2b** (spiropyran → photomerocyanine) reaction ($I'_1{}^{1/2}$, dashed line) compared to data from Experiment I ($I_1{}^{1/2}$, full line).

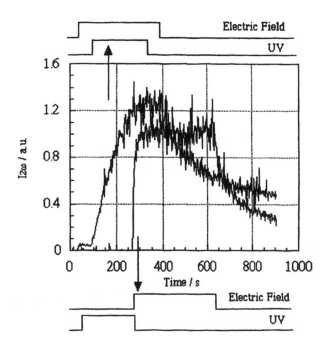

Figure 9 Photoassisted poling at room temperature monitored *in situ* by SHG signal ($I_{2\omega}$) measurement on **3** (furylfulgide-dihydrobenzofuran, 25% w/w).

reaction, **3a** yields a ring closure reaction. When E_p and UV irradiation are both applied simultaneously, the rate limiting process is the creation of **3b** since the kinetics of $I_{2\omega}$'s raise seems to correspond to that of the photochemical reaction. In this system, the poling is "less photo-assisted" than in **2**, since Experiment I and Experiment II show approximately the same SHG efficiency. The maximum d_{33} value reached here is close to 2.5 pm/V in both experiments.

PHOTOSWITCHING OF SHG

Spiropyran-Photomerocyanine (2a–2b)

In photoswitching experiments, polymer samples were previously thermally poled. Despite this poling, these samples show almost no SHG due to weak NLO properties of **2a** (spiropyran). Irradiation with UV light in absence of any external electric field yields an SHG signal that rises up to a plateau value (Figure 10), while **2a** molecules change to **2b** (photomerocyanine). This signal is due to the stronger β value for **2b** compared to that of **2a**. When UV light is switched off, $I_{2\omega}$ slowly decreases, reaching the initial value after a few hours. A subsequent UV irradiation 14 hours later evolves an SHG signal again. This signal revival means that the $I_{2\omega}$ decrease between the two UV irradiation cycles is not due to a total randomization of the molecules. In fact, if this was the only reason of the anihilation of $I_{2\omega}$, the second UV irradiation would not have "revived" SHG. Our observation only means that almost all **2b** molecules faded thermally to their isomer **2a** after the first UV irradiation. However, the maximum value reached after the second UV irradiation is about twice as low as the first one, so a partial disorientation of the molecules can be suspected.

A second type of experiment was performed to an identical polymer film. The sample was alternatively irradiated by UV and visible light, in absence of any external electric field. SHG was followed *in situ* (Figure 11). OD change at 532 nm on another spot of the same sample was recorded to monitor the amount of **2b** under the same irradiation conditions. In the previous experiment, **2b** → **2a** reaction occurs in the timescale of hours because only thermal reaction is involved. Here, visible light accelerates this transformation. In order to be as close as possible to the SHG experiment conditions, we used the experimental setup described in Figure 4 to probe OD, replacing the fundamental beam of the Nd/YAG laser by the second harmonic one. There is a qualitative correlation between the two signals ($I_{2\omega}$ and OD) as they both raise when UV irradiation is switched on, and fall as visible light is applied to the sample. However, the maximum value of OD reached is always the same whereas the maximum value of $I_{2\omega}$ diminishes at each cycle. This difference is obviously due to a partial disorientation when irradiation cycles are repeated. The maximum $I_{2\omega}$ reached in this experiment corresponds to a d_{33} value of 1 pm/V. Another noteworthy fact is that OD does not fall to zero after each visible light irradiation, and that the minimum OD value gets higher and higher. There are at least two possible explanations: the first one is that some **2b** molecules are trapped

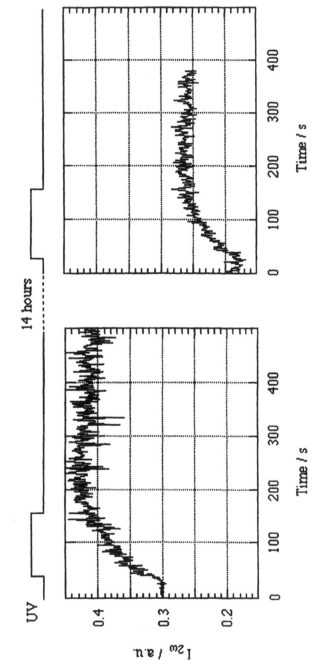

Figure 10 Memory effect in a previously poled **2a** (spiropyran, 25% w/w) film; the subsequent UV irradiation is performed 14 h after the first one.

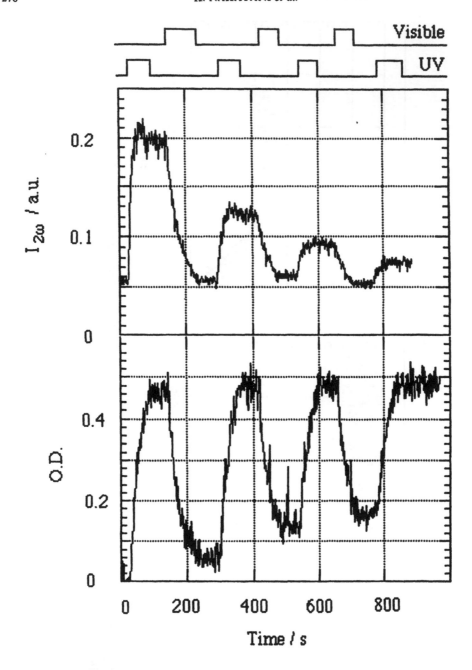

Figure 11 Photoswitching of a previously poled polymer film of **2a** (spiropyran, 25% w/w) film, probed by SHG ($I_{2\omega}$) and OD at 532 nm (proportional to the amount of **2b**) under the same irradiation procedure.

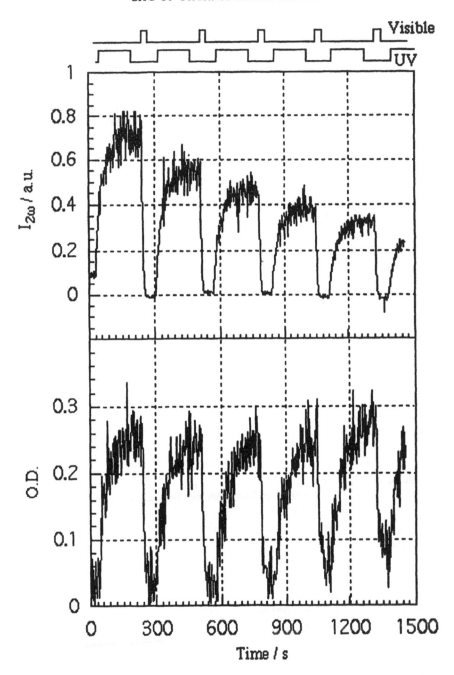

Figure 12 Photoswitching of a previously poled polymer film of **3a** (furylfulgide, 25% w/w) film, probed by SHG ($I_{2\omega}$) and OD at 532 nm (proportional to the amount of **3b**) under the same irradiation procedure.

in small cavities so that they would need more energy to yield **2a**, and the second one is that some **2b** molecules aggregate [26], and lack reversibility.

Furylfulgide-Dihydrobenzofuran (3a–3b)

2a–2b (spiropyran-photomerocyanine) system lacks to fulfill at least two qualities needed to give a good photoswitching device: (i) the thermal stability of the high β species: in the dark, **2b** converts back to **2a** in the timescale of a few hours, and consequently most of $I_{2\omega}$ is lost even if no radiation is applied; (ii) the recycling ability: after a few irradiation cycles, spiropyrans show some fatigue. **3a–3b** (furylfulgide-benzofuran) system may solve these two problems as there is no thermal back reaction from **2b** to **2a**, and as the photoconversion can be repeated several times [16].

As for **2**, a previously poled polymer sample of **3a** was alternatively irradiated with UV and visible laser beams. Here again, an obvious correlation can be established between $I_{2\omega}$ and OD (Figure 12). Moreover, $I_{2\omega}$ has an enhanced reversibility compared to photochrome **2**. This means that disorientation is more important in **2** than in **3**, and a possible explanation is that molecular motion necessary to transform **a** form to **b** form and vice-versa is smaller in the latter photochromic system. Thus, the matrix is less perturbated.

CONCLUSION

An alternative technique to the classical poling method of polymer films (thermally assisted poling) for quadratic NLO has been developed on photochromic polymers. This new method provides several advantages. First, it can be efficiently performed at room temperature. Thus, no heat has to be provided, and we can avoid thermal degradation of the chromophores and thermal reactions during the poling process. It is worthwhile to mention that thermally assisted poling is not suitable for poling **2b** (photomerocyanine) [13]. Second, poling can be performed with a very high resolution compared to thermally assisted poling, by selective and localized irradiation of the polymer sample. This may provide an efficient method to reach quasi-phase matching in waveguide structures.

We also achieved photoswitching of NLO properties of polymer materials. Since light is a convenient way to switch physical properties, signal processing for telecommunication can be performed by this means.

However, after highlighting these phenomena, further investigations are necessary in order to optimize these materials. First, for the photochromic molecule itself, more efficient and suitable photochromes have to be synthesized and studied, in order to avoid photochemical and thermal degradation and, to have a better control of the photoinduced transformation, and finally to have better NLO performances. In this respect, we investigated a few number of spirooxazines [13] and chromenes whose photostability is reputed to be better than spiropyrans. In both cases, no efficient SHG was observed, due to the lack of strong electron-donating or accepting functions, and due to the relatively weak ground state dipole moment of the **b** form.

A better control of the photoinduced transformation means that the most interesting molecules may be either those that do not show any thermal reaction so that nothing can occur unless irradiation is applied or those that yield very fast thermal reaction so that molecular form **b** may be considered as a transient species. Second, for the polymer matrix, the disorientation of molecules in the polymer matrix is a general and major issue for applications of these materials, whether chromophores exhibit photochromic properties or not. This is currently being investigated, and one of the answers to this problem may be the use of rigid high temperature glass transition polymers like polyimides. In the particular case of photochromic chromophores, the polymer matrix is perturbated during the photochemical process. This process acts on the rearrangement of free volumes. Therefore, disorientation may be closely related to this photochemical process.

ACKNOWLEDGEMENTS

We deeply acknowledge Jean-François Delouis who efficiently helped us in performing SHG experiments. We would also like to thank Michel Dumont, Guillerm Froc and Sophie Hosotte for fruitful discussions they inspired and for depositing gold electrodes on thin films, and Isabelle Maltey and Lydia Bonazzola for theoretical calculations.

REFERENCES

[1] D. M. Burland, R. D. Miller and C. A. Walsh, *Chem. Rev.*, **94** (1994) 31.
[2] *Molecular Nonlinear Optics*, edited by J. Zyss (Academic Press, San Diego, California, 1994).
[3] R. E. Schwerzel, *The Spectrum*, Center for Photochemical Sciences at Bowling Green State University, Bowling Green, Ohio, Vol. 6 (1993) 1.
[4] R. C. Bertelson in *Photochromism*, edited by G. H. Brown in Series Techniques of Chemistry, edited by A. Weissberger (Wiley Interscience, New York, 1971), Vol. III, Chap. 3.
[5] R. Guglielmetti in *Photochromism, Molecules and Systems*, edited by H. Dürr and H. Bouas-Laurent (Elsevier, Amsterdam, 1990), Chap. 8.
[6] N. Y. C. Chu in *Photochromism, Molecules and Systems*, edited by H. Dürr and H. Bouas-Laurent (Elsevier, Amsterdam, 1990), Chap. 10.
[7] J. C. Crano, W. S. Kwak and C. N. Welsh in *Applied Photochromic Polymer Systems*, edited by C. B. McArdle (Blackie, Glasgow, 1991), Chap. 2.
[8] V. Krongauz in *Applied Photochromic Polymer Systems*, edited by C. B. McArdle (Blackie, Glasgow, 1991), Chap. 4.
[9] S. L. Gilat, S. H. Kawai and J. M. Lehn, *Mol. Cryst. Liq. Cryst.*, **246** (1994) 323.
[10] L. Läsker, T. Fischer, J. Stumpe, S. Kostromin, S. Ivanov, V. Shibaev and R. Ruhmann, *Mol. Cryst. Liq. Cryst.*, **246** (1994) 347.
[11] K. Ichimura, *Langmuir*, **11** (1995) 2341.
[12] R. Loucif-Saïbi, K. Nakatani, J. A. Delaire, M. Dumont and Z. Sekkat, *Chem. Mater.*, **5** (1993) 229.

[13] Y. Atassi, J. A. Delaire and K. Nakatani, *J. Phys. Chem.*, **99** (1995) 16320.
[14] H. Rau, G. Greiner, G. Gauglitz and H. Meier, *J. Phys. Chem.*, **94** (1990) 9523.
[15] E. Fischer, *J. Phys. Chem.*, **71** (1967) 3704.
[16] H. G. Heller and J. R. Langan, *J. Chem. Soc. Perkin II* (1981) 341.
[17] B. F. Levine, C. G. Bethea, E. Wassermann and L. Leenders, *J. Chem. Phys.*, **68** (1978) 5042.
[18] M. Dumont, Z. Sekkat, R. Loucif-Saïbi, K. Nakatani and J. A. Delaire, *Nonlinear Optics*, **5** (1993) 395.
[19] P. M. Blanchard and G. R. Mitchell, *Appl. Phys. Lett.*, **63** (1993) 2038.
[20] Z. Sekkat and M. Dumont, *Appl. Phys. B*, **54** (1992) 486.
[21] Z. Sekkat and M. Dumont, *Nonlinear Optics*, **5** (1992) 359.
[22] J. A. Delaire, Y. Atassi, R. Loucif-Saïbi and K. Nakatani, *Nonlinear Optics*, **9** (1995) 317.
[23] M. Dumont, G. Froc and S. Hosotte, *Nonlinear Optics*, **9** (1995) 327.
[24] G. Williams and D. C. Watts, *Trans. Faraday Soc.*, **66** (1970) 80.
[25] H. L. Hampsch, J. Yang, G. K. Wong and J. M. Torkelson, *Polymer Comm.*, **30** (1989) 40.
[26] J. B. Flannery, *J. Am. Chem. Soc.*, **90** (1968) 5660.

INDEX